ARE YOU THERE GOD?
It's me, Darwin.

by Jacob Johnson

The social instincts ... with the aid of intellectual powers and effects of habit, naturally lead to the golden rule, "As ye would that men should do to you, do ye to them likewise."
— Charles Darwin, 1871

For my family.

Published in the United States by Informing to Empower, an imprint of
Progressive Source Publishing, California

Illustrations by Julie McDowell
Cover Art by Rachel Hudson
Back cover design by Djurica Bogosavljev
Formatted for print e-book and audio by Kai Daniel Lewis

Print ISBN: 978-1-7353785-2-7

Table of Contents

Prologue

In the year 1832, a young man from the largest naval empire in history stands on the deck of a wooden survey ship and gazes up at the stars. All around, shining brightly above the dark and churning ocean, burns undeniable proof of the Creator and His works.

Charles Darwin, a recent graduate of Cambridge's prestigious Christ's College and soon-to-be clergyman, is bound for the rainforest. He will soon see life climbing, soaring, radiating out into the world in so many forms that his enormous collection of textbooks cannot classify them all. The world is glorious, and God's wonders lie all around, if one only knows where to look.

Someday, decades later, an aging Darwin will look up at those very same stars and wonder where the Creator has gone. His memory will burn with thoughts of his dead daughter Annie, slaves screaming in agony, and nightmarish parasitoid insects. Why might life be so beautiful, yet so terrible? Why might we be made so loving, and yet so hateful? Is He even listening? Is He even there?

But on this night, the man is still young. On this night, he breathes in the sea breeze and feels the wind in his hair. On this night, before sleep takes him, Charles Darwin closes his eyes and begins to pray.

///

This is a book about the development of human morality and the evolution of Charles Darwin.

Today, science and religion are so often pitted against each other that they've lost sight of their moral common ground. As humanity faces existential crises like mass extinction and environmental catastrophe, we can learn a lot from Darwin. The man whose initial work now fuels hostility between faith and science not only struggled mightily with the question of God, but also with the question of good and evil.

When *On the Origin of Species by Means of Natural Selection* released in 1859, many people of faith decried it as a fundamentally immoral book. Evolution was seen as cold, dispassionate, and deprived of the loving guidance of a Creator. But those who thought that Darwin's biological theories destroyed the foundations of morality were wrong. In fact, evolution *made those foundations stronger.*

In *The Descent of Man, and Selection in Relation to Sex*, Darwin's book on human evolution, he posed the radical idea that humans developed a moral sense of right and wrong from evolutionary processes themselves. He asserted that humans - with all of our complex moral and spiritual reasoning- evolved right along with God's other creations. But despite becoming one of the most controversial books of all time, *The Descent of Man* was never intended to undermine religious beliefs.

A century and a half after Darwin's monumental work, the field of evolutionary biology is nearly unrecognizable compared to what it was in the late 1800s. Biologists spent the decades after Darwin poring over the natural world. Through observation and experimentation, they found the planet Darwin saw — a place filled with immense beauty and profound strangeness. Within the wondrous study of life, people have continued to search for the origins of "good," "evil," and the moral systems we judge them by.

By studying the behavior of other animals, humans are better able to understand their own. Social and cultural behavior– as it turns out– is animal behavior, too. Some philosophers are delving back into our biological lineage for answers to troubling ethical questions about the future of the human species. Humans are struggling to change their collective behavior as they threaten the biological world and themselves with it.

This book revisits Darwin's theory of how morality evolved, equipped with scientific knowledge that was unavailable to the father of natural selection. By exploring some of the strangest behaviors that the natural world has to offer, we can better understand the origins of human morality, which are shaped by the same evolutionary pressures that make birds and bees. And we can even ask that age-old question:

Are you there, God?

Introduction

When I was in third grade, my great-aunt Eleanor, a devoted Christian, asked me an important question:

"You're not one of those people that believes in evolution, are you?"

I didn't know how to respond. I was, indeed, one of those people. But from her tone and the way she scrutinized me from across the table at a diner in Muskegon, Michigan, my eight-year-old self decided to not give a straight answer. I mumbled something about the mysteries of the universe and returned to my crayon-covered crossword.

Then, in 2014, during my junior year of high school at a mock government camp called Boy's State, my mock science funding bill was rejected as one of my fellow "senators" got up and exclaimed: "The Bible tells us that we didn't come from monkeys!" Causing our debate to rapidly *devolve*, if you will, into an argument over the truthfulness of science.

From that moment on, I became fascinated by the dissonance people experience between science and religion. Why did the two seem irreconcilable? I started reading the Bibles I found in hotel rooms. I attended Bible study in college, and talked to preachers on street corners. Over the years, one name came up again and again: *Darwin.*

Darwin, I was told, was an atheist. In rejecting God, Darwin, it was said, undermined the foundation of all human morality because his vision portrayed us as soulless. Lost. Doomed. Darwinism, many tell me, is antithetical to Christianity. Darwinism will take you to hell.

It is 2020. I'm about to start a PhD in animal behavior, the COVID crisis is in full swing and the world is going to hell. I am, like many others, in a kind of semi-permanent foul mood. In prior years, I've helped graduate students out with biology research, studying the ecology of birds in countries like Denmark and Peru. My field season this year has just been canceled. Cooped up at home and unable to look away from the news, I stew and angrily ruminate about viruses and climate change and the evils of human nature. My stolen hotel room bible tells me that when God finished creating the world, he saw that it was good. Good? The more I look at our world, the more it seems like a brutal place, interspersed with cute animals.

An article pings on my email– something about a cute animal, perhaps? Nope. It's about wasps, and some nasty ones, at that. Researchers are starting to call parasitoid wasps "Darwin's Wasps." The reason for their namesake? Because these wasps have such violent life-cycles that they made Charles Darwin begin to doubt the existence of a benevolent God.

In Peru, I'd worked with a biologist who'd had his own run-in with a Darwin's Wasp. He had been deep in the Amazon jungle and got stung in the leg by one. He told me that these wasps lay eggs inside of their stung host's body, which hatch underneath the skin, feed on their insides, then burrow out once they reach adulthood. In other words, he was playing host to a larva in his leg. Gross.

"It's really cool," the biologist had reassured me with a grin. "Want to see?"

"I'm good, thanks." I'd replied. I, too, doubted the existence of a benevolent God after receiving this information. Maybe instead, they should have named it the 'Atheist's Wasp.'

That miniscule creature buried in the researcher's leg had, I mused, made Charles Darwin one of the most prominent nonbelievers of them all. But... was it really that simple? Or was this just the wasp that broke the naturalist's back?

Then it hit me. I'd spent so much time learning about evolution and talking about the compatibility between evolution and religion, but I never knew what Darwin believed in when it came to God, or why he held whatever beliefs he did. Sure, people threw Darwin's name around in debates with Creationists or lectures about human progress, but how often did they talk about what Darwin actually *thought?*

I began reading books about Darwin, only to discover that everything I had assumed about his faith and views on morality was wrong. Darwin was a Christian when he first developed evolutionary theory. His instructors in natural history were important members of the Christian clergy. And scarcely anybody - not the preachers, not the scientists - ever mentions it.

My Aunt Eleanor passed away not long after the conversation in that diner, so giving her this book is unfortunately off the table. Some of my fellow Boy's State senators are probably thinking about running for *real* congress soon. If one of them happens to read this, they might be curious to know that before Darwin was a scientist, he was a priest-in-training. And before he was a priest-in-training, he was a doctor-in-training. He enjoyed being a priest-in-training. As a doctor-in-training, he was thoroughly miserable.

Part I: Morality and the Tree of Life

1. Building Moral Compasses

"A moral being is one who is capable of comparing his past and future actions or motives, and of approving or disapproving of them. We have no reason to suppose that any of the lower animals have this capacity."
— Charles Darwin, *The Descent of Man*

It is loud in the surgery room. Anesthetics are uncommon in this day and age. Charles steals a glance at the grisly scene, then grips down on his notebook so hard that the bindings impress into his palm. He doesn't want to be here.

As a medical student at the University of Edinburgh, Charles is required to sit in at a live surgery and take notes. The person on the operating table is a child. He spots a cringe on one of the other students' faces as another cry of pain is let out.

Medical school has gone from utter boredom at the droning medical lectures and tedious anatomy to sheer terror in an instant. The pen trembles, making Charles' already messy handwriting borderline illegible. He wishes he were out collecting beetles, or messing around with a chemistry set with his brother Ras, or even back in boarding school. But he can't go back. He has a legacy to live up to.

Both of Charles' grandfathers are physicians. One of them– Erasmus Darwin– was invited to be royal physician to King George III. Charles' father, Robert Darwin, is a highly successful medical doctor with a large practice. Ras is well on his way to becoming a doctor, too.

Meanwhile, Charles has never been a model student. He's taken up naturalism as a hobby, but serious subjects like law or medicine have never clicked. His middling grades caused his father to tell him "You care for nothing but shooting, dogs, and rat-catching, and you will be a disgrace to yourself and all your family."

Charles does not want to be a disgrace to his family. The morally right course of action is to follow their wishes– to be a good son. That's why he's here in this bloody surgery room.

There is a lot of blood. He imagines what kind of pain the poor child must be in. Charles' stomach turns over and he fights every impulse to get as far away from this place as possible. "Honor thy father and mother" is a Commandment, but is it morally right for him to endure a lifetime more of this? Trapped in a profession that he despises just because his father wants him there?

Maybe he could become a priest.

///

Across centuries and civilizations, humans agree: we are different from other animals. When asked to define the difference, most people say something like "intelligence." Philosophers might say "morality." Theologians might say "belief in God." These are tricky terms to define, especially in the context of other animals. Human uniqueness stands on some very shaky ideological foundations.

Take "intelligence." Nobody can agree on a consistent set of criteria, not even among humans. IQ tests are scientifically problematic; they don't represent all of a person's cognitive abilities. If it doesn't work well for humans, what do we mean by "intelligent" in the context of other animals? Are plants morons because they lack emotional intelligence? They don't even have brains, let alone emotions. Why would a solitary polar bear even need social intelligence the way we do? When humans judge other creatures using themselves as the gold standard, bad things happen. In fact, bad things can happen when humans try to judge *anyone's* intelligence. The judges aren't half as smart as they think they are.

What about "consciousness?" More problems. We can't know exactly what other animals are experiencing without some kind of sci-fi mind-reading machine. We're not even sure what consciousness means for *us*. What about the last two features that people use to set humans apart — morality and belief in God? Do other animals have morality? Can other animals believe in a god?

Morality means a set of principles used to distinguish right and wrong, good and bad. A good act is "moral," and a bad or evil act is "immoral." Humans are moral animals. Is there evidence that other animals know whether they have done something right or wrong? When your dog rips apart the couch cushions and gives you a sad look afterward, whimpering when you turn on the lights and see the mess, does it feel guilty? Does it know it did something wrong? We can't completely "know the mind of dog" yet, but most experts don't think so. Primatologist Christopher Boehm tells us

> It is with a wistful sense of personal regret that I must inform other loyal dog owners that when their charges *seem* to be giving them guilty looks, in all probability they (the owners) are projecting their own moralistic human reactions onto an amoral canine. An empathetic dog may be feeling uncomfortable in the face of disapproval, or submissively fearful of punishment for breaking a rule ... but I'm reasonably certain that a

> humanlike sense of being *ashamed* ... plays no part in this picture.

So maybe, to the dog, "right" and "wrong" are governed by whether you react positively or negatively to the things it does. That's not morality.

Dogs have evolved to be social, cooperative animals. Like us, their wolf ancestors hunt in packs and have a social hierarchy. It makes sense that dogs react so emotionally to our approval or disappointment — we're the ones in charge. Hopefully.

Cats, on the other hand, respond very differently to your disappointment than dogs do. Cats will hear you berating them about bringing in a dead squirrel and shoot you a blank stare of utter apathy. They couldn't care less. This is probably because the felines that house cats descend from were solitary predators — no pack hunting, no complex social hierarchy. The social "pre-moral" instincts of dominant and submissive pack behavior in dogs aren't really there in cats. That's why cats are a lot better at being worshiped as gods than they are at fetching your newspaper.

Speaking of gods, lots of dogs seem to view their masters that way. One of Darwin's contemporaries, William Braubach, even wrote as much in a response to Darwin's work. Humans give Fido his daily dog food, pass down commandments ("thou shalt not relieve thyself indoors"), and deal out punishments and rewards for bad and good behavior. Many people believe that God punishes the wicked in hell and rewards the faithful with blessings and eternal life. Is the Christian God like an omnipotent dog owner or shepherd, making us to lie down in green pastures and coaxing us to do the "moral" thing with the promise of a reward?

Above: a human, a dog, and a cat. These three species all have very different conceptualizations of good and bad behavior, but they seem to get along well enough. Photo by Lissák Tivadar. Accessed via Wikimedia Commons.

"No," the wannabe philosopher in my mind protests. "That's not morality. Morality means conceiving of standards that go beyond punishment and reward. It's about holding fast to these principles and using them to make tough choices. It's about acting virtuously in the face of sheer adversity."

"No!" adds a hypothetical theologian. "God isn't just a master telling you right and wrong. He gave you free will to make your own decisions, and it's up to you to choose the right path. He created the universe. He's so much more than any animal could imagine."

There isn't much to be gained from nitpicking at definitions in order to make ourselves distinct from the rest of the animal kingdom, so let's try a different question: where does our human morality and belief in God *come from* in the first place?

The origin of morality was a big question in Darwin's *The Descent of Man*. Its thesis is that humans, with all their behavioral complexity, arose just like any other species from natural and sexual selection. For many thinkers at the time, morality was so unique, so exclusive to humans, that it was doubtlessly "God-breathed," i.e. actively given, by a benevolent creator, to humanity, which religions like Christianity suggest God made in His own image.

Darwin didn't buy this logic by the time he published *On the Origin of Species*. Darwin maintained that, even though it may seem impossible that an ant could evolve over time without a designer, that's in fact what happens through tiny changes over millions of years. The problem is our perception of time.

Human lifetimes last a century at best. Life on Earth has been around for 3.6 billion years, which is about 36 million very long human lifetimes. We can't even *fathom* 3.6 *billion* years. Millions of large stars in our galaxy formed, lived, and died over that period. On this kind of scale, mountains move, continents split, and complex life can be naturally selected for. Darwin believed that morality developed at a similarly gradual, albeit faster pace than the evolution of something like the eye. According to *The Descent of Man,* morality began with simple principles. It then developed more complications, reaching higher and higher echelons of development. Each step along the way was like a rung on a ladder, leading from basic care of family to tribal unity to the Golden Rule itself.

So, where do dogs rest on the ladder of morality? When trained by humans, dogs can resist temptation in order to follow rules set by their masters. Though your dog might instinctively want to grab that squirrel, it may resist that urge, because it roughly understands that breaking loose is "being a bad dog." Or at least that treats may be withheld. This kind of "premoral" behavior is rudimentary, only setting the groundwork for a moral compass.

Tiny, social insects like ants were a massive problem for the budding theory of natural selection. Most ants don't reproduce and will actively lay down their lives for other members of the colony. Worker bees' stingers rip out of their bodies and kill them in the process. What evolutionary benefit did most bees or ants get from colony life? After months of befuddlement, Darwin realized that natural selection could still work on individuals to benefit the colony as a whole. Its members are all *related*. The evolutionary justification for a honeybees' kamikaze sting is to promote the genes of relatives — the longevity of their own DNA.

Human behavior is tricky. To make his point about the natural progression of morality in *Homo sapiens*, Darwin compared his British culture to those of "savages." He meant indigenous people from parts of Africa, Asia, and the Americas, whom Darwin viewed as less socially developed than people from his motherland of Britain. Young Darwin considered the country that eats baked beans for breakfast as the pinnacle of civilization.

Darwin saw the differences in human cultural and technological "progress" as evidence that morality could go through stages of development. As if reaching toward greater and greater complexity. It is important to understand that Darwin was a product of his nineteenth century environment. English "cultural superiority" was seen as scientific fact by many baked bean-eating Victorian gentlemen. Darwin may have been a Victorian, but he was still an unusually open-minded and insightful one. He frequently wrote about the horror of slavery, even changing his views on native Fuegians later in life. Other examples support the point that Darwin was trying to make. Most laws need moral assumptions (such as "stealing is wrong") to make sense. Roughly four thousand years ago, the ancient Babylonian king Hammurabi was said to have created a code of laws– one of the earliest written "legal systems" that we know of. Its many rules are inscribed in Akkadian atop a black stone pillar, detailing punishments such as "if a man destroys the eye of another man, they shall destroy his eye." However, a lot of the Code's punishments are death. Theft? Death. Adultery? Death. Lying in a capital trial? Building a house that later falls down? Unfortunately, still death.

Hammurabi's Code was probably effective when it was written. As societies developed, legal systems across the world changed. Many countries now try to avoid cruel and "unusual" punishments. Cutting out someone's eye generally qualifies as cruel and unusual today. Humans increasingly oppose the death penalty though, just a few centuries back, a fun English afternoon activity was watching someone get their head cut off. Moral ideas of how to deal with crime and punishment have become more nuanced, and as Darwin would argue, more developed. Humans have progressively realized that "an eye for an eye makes the whole world blind."

Above: Hammurabi's Code. Its mortal punishments made breaking the law inadvisable. Photo by Charlie Phillips. Accessed via Wikimedia Commons.

Even if morals do not improve over time, they certainly change — or "evolve," in a manner of speaking. Evolution doesn't actually imply "advancement," rather just "descent with modification." Darwin's ultimate point is that even the earliest human society would still be miles ahead of any other social animal's sense or system of morality, and yet we can still see clear signs of basic organizational principles in social animals, including emotions and empathy. Is it that big a stretch to suggest that morals have their roots in the evolution of social species?

In *The Descent of Man,* Charles argues that social organisms underwent natural selection for behaviors encouraging cooperation and co-habitability. These behaviors might include sharing resources, caring for sick individuals, and your roommates remembering to do the dishes.

There are many benefits to living in a group, like protection from predators, shared care of young, and easy access to mates. There are also many drawbacks, such as disease or the dangers posed by jealous and hungry group members. If group members compete too aggressively or too often, the benefits of group living get outweighed, and most individuals will be better off on their own.

Darwin reasoned that certain basic instincts that encourage group compatibility would be selected for. Eventually that selection, combined with high cognitive abilities, might lead to a heightened sense of right and wrong. He writes:

> The following proposition seems to me in a high degree probable — namely, that any animal whatever, endowed with well marked social instincts, the parental and filial affections being here included, would inevitably acquire a moral sense or conscience, as soon as its intellectual powers had become as well, or nearly as well developed, as in man.

Social instincts set the groundwork for the development of morality. For Darwin, one of the most important of these instincts was "sympathy," or the capacity to feel the pleasure or pain of others. Darwin partly agreed with contemporaries who proposed that feelings of sympathy arise when the experiences of others bring forth memories of our own experiences, writing:

> We are thus impelled to relieve the sufferings of another, in order that our own painful feelings may be at the same time relieved. In like manner we are led to participate in the pleasures of others. But I cannot see

> how this view explains the fact that sympathy is excited, in an immeasurably stronger degree, by a beloved, than by an indifferent person. The explanation may lie in the fact that, with all animals, sympathy is directed solely towards the members of the same community, and therefore towards known, and more or less beloved members, but not to all the individuals of the same species.

Darwin was right to point out that individual empathy should only extend so far. It's not very helpful if you're in a wolfpack and start to sympathize with the deer you're chasing down.[1] On the other hand, if you're sensitive to the feelings of your pack, you can better read their intentions and emotions, better coordinate your hunts, and function as a more cohesive group overall. All for one and one for all. Darwin concluded:

> We are led by the hope of receiving good in return to perform acts of sympathetic kindness to others; and sympathy is much strengthened by habit. In however complex a manner this feeling may have originated, as it is one of high importance to all those animals which aid and defend one another, it will have been increased through natural selection; for those communities, which included the greatest number of the most sympathetic members, would flourish best, and rear the greatest, number of offspring.

[1] Stay tuned, though: early humans may have used empathy as a hunting tool.

Mutualism! I scratch your back, you scratch mine. The more metaphorical back-scratching, the better. There's one more leap we have to make from mutualism to higher ideas of morality. Well, more like a crawl.

It's a gradual process. Social instincts give an early human "some wish to aid his fellows, some feeling of sympathy ... Such impulses will have served him at a very early period as a rude rule of right and wrong." From there,

> he regarded more and more, not only the welfare, but the happiness of his fellow-men; as from habit, following on beneficial experience, instruction and example, his sympathies became more tender and widely diffused, extending to men of all races, to the imbecile, maimed, and other useless members of society, and finally to the lower animals — so would the standard of his morality rise higher and higher.

This was a revolutionary idea for its time. Darwin's work struck at the heart of the moral absolutes that philosophers held dear, bridging the gap between humans and the rest of animal life in a way that few other evolutionary ideas could. The moral spark that humans use to separate themselves from all other species did not just appear out of the blue or on a set of tablets. It grew and developed out of something more primordial, more universal. The idea of "emergent properties" comes to mind, where a combination of smaller, simpler things leads to something — a process, an organism, etc. — that is greater than the sum of its parts. A helium atom is functionally more than just two protons, two neutrons, and two electrons. A human is functionally more than just 37 trillion cells. We could become greater than the sum of *our* parts, claimed Darwin, using our own moral sense of right and wrong.

Of course, moral ideas do not necessarily come from selection for socialization. Nature and nurture both have roles to play:

> It is, however, impossible to decide in many cases whether certain social instincts have been acquired through natural selection, or are the indirect result of other instincts and faculties ... or again, whether they are simply the result of long-continued habit.

Darwin still warned about accepting "just so" explanations for anything as complex as human morality. The brain is one complicated piece of biological machinery. Not all behaviors are the result of adaptation. Still, the social behaviors described by Darwin, like group defense and alarm calling, appear in unrelated social species across the animal kingdom. Natural selection plays a big role in the selection of certain social behaviors.

Moral development is not just from, as Darwin puts it, "a sense of dissatisfaction, shame, repentance, or remorse ... after having yielded to some temptation." It arises from the ability to extend care beyond ourselves, to our families, to entire communities around us. As Darwin argues, morality plants its roots in a higher concept: the common good.

> The prime principle of man's moral constitution — with the aid of active intellectual powers and the effects of habit, naturally lead to the golden rule, 'As ye would

> that men should do to you, do ye to them likewise;' and this lies at the foundation of morality.[2]

For a man whose work is so often pitted against religious ideas, this statement may seem surprising. We've arrived at the crux of Darwin's argument: The Golden Rule, and more broadly, the Moral Compass, developed naturally over evolutionary time. Darwin doesn't remove God from the picture, though. But where is He, then? Hiding amongst the Moral Compass' many pieces? With the help of modern-day biology, we'll search step by step, piece by piece, for how exactly we've come to know good and evil.

---

[2] The Golden Rule isn't exclusive to the New Testament. Confucius wrote in 500 BCE: "What you do not want others to do to you, do not do to others." In 375 BCE, Isocrates wrote, "Do not do to others what would anger you if done to you by others." These statements are inversions of the Golden Rule, sometimes called the Silver Rule. Two hundred years later, the *Mahabharata* reads: "This is the sum of all true righteousness: deal with others as thou wouldst thyself be dealt by. Do nothing to thy neighbor which thou wouldst not have him do to thee hereafter."

2. Cannibalism: A Crash Main Course

"Derogatory that the Creator of countless systems of worlds should have created each of the myriads of creeping parasites and worms which have swarmed each day of life on land and water on this one globe"
— Charles Darwin, personal notes

As we investigate the evolution of morality in living creatures, let's re-address the parasitoid wasp in the room.

The year is 1860, and Charles Darwin has just penned his famous letter to Asa Gray referencing the *Ichneumonidae*, a genus of parasitoid wasp. Why did Darwin refer to these wasps in a letter about doubting the existence of a benevolent God? To start, a parasit*oid* is a little different from a parasite. While a parasite is happy to live off a host without killing it, a parasitoid aims to kill its host. Their tactics are as diverse as they are gruesome.

Parasitoid wasps usually pick a specific host like caterpillars or cockroaches. Consequently, these wasps have diversified into various groups that each specialize in a specific host.. One group– the *Ichneumonidae*, aptly nicknamed Darwin's Wasps, contain around 100,000 species. That's nearly twice as many species as all of the fish, mammals, birds, reptiles, and amphibians combined. These creatures are just about everywhere.

Darwin's Wasps all share a similar reproductive cycle. The female wasp stings her prey, lays eggs in the host, and departs. The eggs eventually hatch, and the larvae begin to consume the host from the inside out, politely saving the vital organs for the finale to ensure the host doesn't die and decay too soon. Finally, the newly grown wasps emerge from the host, leaving their victim a decrepit husk.

That's nature for you. Most hunting species show little regard for the suffering of their prey, and even "high and mighty" humans can behave like that. On Earth there are plenty of parasitic species, or species that cannibalize one another, or species that take over the minds of others. Sometimes, species have antagonistic relationships, full stop. Not always. Sometimes, when we take a closer look at even the most bizarre and unsettling behaviors, we can find the driving forces behind the evolution of morality hard at work.

Cannibalism runs rampant in the animal kingdom. Most humans would consider this behavior reprehensible, except maybe under extreme circumstances. In *Cannibalism: A Perfectly Natural History*, zoologist and science writer Bill Schutt defined cannibalism as "The act of one individual of a species consuming all or part of another individual of the same species." We'll use that same definition here.

Cannibalism occurs in insects, crustaceans, fish, amphibians, mammals, and other animals. The most common form is filial cannibalism: eating one's own young. I think we can all agree that eating people is generally bad, killing infants is very bad, and eating newborns is downright atrocious. In other species, though, there can be good reasons for it.

As any human parent knows, children are expensive, exhausting, and time-consuming. Parents need nutrients to offset this cost, and organisms sometimes draw them from the nearest source available: their offspring. This sounds paradoxical. Reproduction requires a lot of energy, too, so why on Earth would an organism eat its own offspring? Sometimes, they just have too many kids.

Biologists weigh the differences between "K" and "R" reproductive strategies[3] in the animal kingdom. An "R" strategist invests little energy in each individual kid, but creates many to compensate. Most insects are "R" strategists. Their lifespans are short, and the survival rate of their offspring is low, but if a few, or even just one, of the original batch of kiddos survives to adulthood and reproduction, the parents will have done their "jobs". In contrast, "K" strategists like large mammals invest a lot more energy into each of their offspring, and therefore have fewer of them. The Ks are longer-lived and have higher survival rates than their R counterparts. A typical fish or crab (both of which have many young, but relatively few that survive to adulthood) is usually considered an R strategist. A whale or an elephant, both of which carry and care for their offspring for many years, are solid K strategists. Humans, too, with our nine-month gestation and absurdly long childhood, fall into the K strategist category.

---

[3] The "K" and "R" refer to "carrying capacity" and "rate of exponential growth" on a graph.

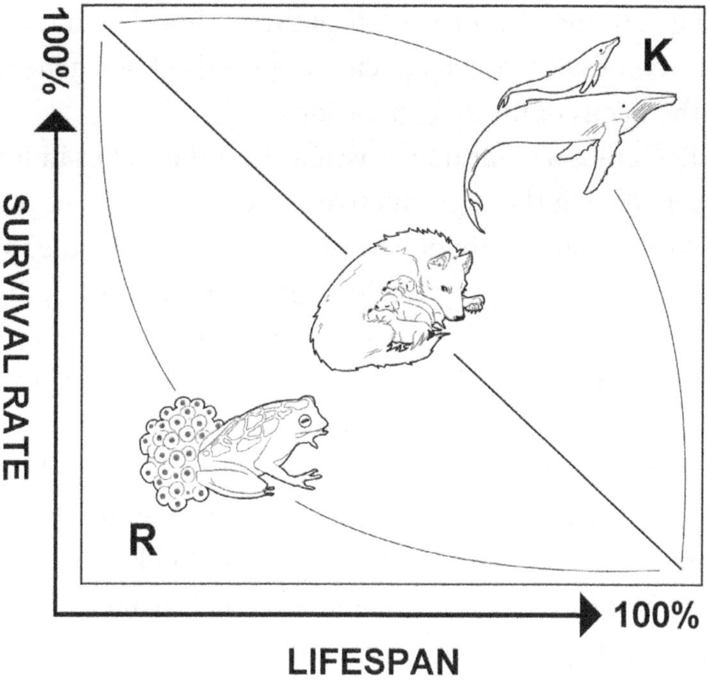

Above: "K" and "R" strategists. R strategists invest a little into many offspring, K strategists invest a lot into very few offspring.

It's no surprise that animals like whales or elephants never cannibalize their children while fish and crustaceans do so all the time. The difference in energy investment drives the differences in behavior. There are always exceptions to this rule. For instance, the male polar bear (a K strategist) will often eat other males' cubs, not just for nutrition in a nutrition-scarce environment, but to bring female bears back into estrus (or mating shape). Likewise, some mice (generally R strategists) will cannibalize some of their babies to mitigate overcrowding in the nest and provide more resources for their other babies.[4] Barring extreme instances such as these, though, cannibalism usually comes down to balancing the needs of one generation with the needs of future generations.

Red crabs appear on Christmas Island beaches in huge numbers during the reproductive season. Once the eggs hatch, the adults wander through swarms of their own young, picking babies up with a careful crimson pincer and devouring them. It makes sense; the vast majority of these young crabs aren't going to survive anyway, so the mom might as well try and get some nutrients back before starting her next brood.

Some cichlids, a kind of fish, carry their eggs and young around in their mouths for safekeeping, an adorable practice that makes its way into documentaries and kids movies like *Finding Nemo*. What these media often fail to mention, as Schutt rightly points out, is that since carrying young in your mouth makes eating difficult, cichlids feed themselves during this phase by swallowing some of their eggs. If her brood gets too low, the cichlid may simply swallow the whole lot and start over. That's the way it goes. The young of any species tend to be ideal targets for cannibalistic behavior because they are both defenseless and readily available. The offspring of R strategists live hard-knock lives. I doubt the parents feel much guilt over it.

---

[4] In general, cannibalism can work as a response to resource scarcity not only because the parent gets nutrients, but also because this reduces crowding and competition for the resources currently available.

Above: a chichlid's young peer out from within their parent's mouth. Photo by Matthew Miller. Accessed via Wikimedia Commons.

Siblings can be more than rivals. Some animals, like spadefoot toads and sand tiger sharks, practice sibling cannibalism. Sand tiger pups do this in-utero, with young sharks turning on one another for food when their yolk begins to run dry. In the end, only two sharks remain, one in each oviduct. Some researchers suggest that this behavior feeds the strongest offspring and begins to teach them useful hunting skills. It's a shark-eat-shark world from the get-go.

Spadefoot toads eat one another largely out of necessity, though not all tadpoles turn to cannibalism. Spadefoot toads are desert specialists that need bodies of water for their young to develop. They solve this challenge with patience. The toads wait underground for the rare desert rains, then quickly emerge at the surface and lay their eggs in ephemeral pools. The tadpoles that hatch must develop quickly, before the pools dry. Some manage to survive on plankton and organic debris, but the more successful tadpoles usually turn to eating the competition as quickly as possible. It's a perfectly functional strategy, and the tadpoles even discriminate, leaving some siblings uneaten. They're selective, likely based on kin selection, which we'll discuss more later. For those who make it through birth and grow to adulthood without being cannibalized, a new lethal danger waits: sex.

3. Femme Fatale

"Charms of music & female chit-chat. — These things good for one's health. — but terrible loss of time."
— Charles Darwin, personal notes.

Perhaps you've heard of female praying mantises eating their lovers' heads after, or even during, the act of copulation. While that kind of sexual cannibalism happens under laboratory conditions where the female is starving, the act is much more common in species like octopuses and spiders. One Australian species of spider in particular has piqued the interest of researchers on account of the absurd *effectiveness* of its postcoital performance.

The redback spider demonstrates the sheer power of sexual selection. Redback spiders, like most spiders, are highly sexually dimorphic, meaning that two sexes have markedly different physical characteristics. In this case, the female redback spider is much bigger than the male, so much so that he could wind up as the "dinner" portion of a dinner date if he tries to seduce her when she's hungry. Some animals with similar sexual dimorphism take steps to avoid being eaten after sex. For instance, many male octopuses use a "mating arm" to stay as far away from the female as possible while delivering their sperm, because after mating is over, the risk of death is, well, pretty high. So great is the danger of strangulation and cannibalism to the unwary male octopus that some will detach their penises and leave them for the females to use rather than stick around themselves. Other species jet away as quickly as possible after the deed is done, before the female realizes she could go for a snack.

The male redback spider does none of these things. Instead, after sex, he performs an Olympian somersault directly into the female's waiting jaws, offering himself up as food. Why would a male capable of mating multiple times throw that all away on a first date? This apparently counterproductive behavior was resolved by researchers led by Dr. Maydianne Andrade from the University of Toronto, who found that the male redback spiders who let themselves get eaten not only sired more offspring, but also kept their partners from seeking extra mating encounters with other males. Bizarrely, the combination of female preference and the bonus nutrition provided by the male is so strong that post-coital death somersaults are evolutionarily advantageous.

Rather than the male spider's ensuring its individual success, he deliberately gives his life for the success of his progeny. This relentless push to pass one's genes to the next generation can manifest itself in cruel ways, but I see something noble–even admirable–in species that sacrifice themselves like this. And out of all the ways to go, a post-coital death somersault seems like a pretty damn awesome one.

Another species of spider gets cannibalized by its kids. The female black lace-weaver spider, unable to keep feeding her young with her extra unfertilized trophic eggs — basically pre-packaged baby food — eventually calls her young to her by plucking on her web, offering her body up as nourishment. Caecilians (a kind of legless amphibian) feed their young by letting them bite off pieces of their own skin, which seems like a much more painful alternative to breastfeeding.[5]

There are, of course, cases where cannibalism seems to lack strategic value, but instead manifests itself as the by-product of other behaviors. Chimpanzees killing and cannibalizing an infant, chickens pecking at one another, and Hannibal Lecter are all examples of cannibalism that does not play an important role in the species' survival. Humans are the only living species known to practice the occasional form of cultural, ritualized cannibalism as a means of honoring the dead, but this practice deserves its own category. These instances of "less useful" cannibalism are rare in nature, relative to more "rational" or self-sacrificial forms.

Past generations go to admirable lengths to produce the next, even if that involves a little cannibalistic infanticide from time to time. Species practicing filial cannibalism normally do it in order to be healthy enough to make another generation; the end goal is the same, even if the behavior changes dramatically. Seems there's more than one way to skin a caecilian.

In some instances cannibalism can be good, even necessary, for group cohesion. One species has adopted it as an admittedly unconventional way of ensuring fair play. This next section is for the birds.

---

[5] To be exact, the young only eat a specialized nutritional upper layer of the skin, but still.

4. Bird Eaters, Bird Feeders

> "What a book a Devil's chaplain might write on the clumsy, wasteful, blundering low and horridly cruel works of nature!"
>
> — Charles Darwin, from a letter to Asa Grey

After two years, Charles has dropped out of medical school in Edinburgh. This comes as a disappointment to his father, but all is not lost. There are still a few other professions that would be respectable for one of Charles' family's station, like "military officer" or "clergyman." They go with the latter, and young Darwin is quickly enrolled at Cambridge to pursue a Bachelor of Arts on his way to the priesthood. In this age of keeping up appearances among kin, Darwin's father's motivations are twofold. He wants to maintain the good family name, yes, but Robert also wants to keep his son from being idle– to be able to support himself and eventually a family of his own. Charles' concerns are far from these lofty ideals, though. College is going to be fun. He is finally away from all of those boring anatomy lectures and screaming patients. Here, he can let his mind free.

Charles meets Reverend John Stevens Henslow– an experienced naturalist in his own right– as a student in Henslow's botany course. Henslow's teaching style clicks with the young scholar, reigniting a passion for the natural world that has been with Charles since early childhood. Teacher and pupil become good friends. Thanks to Henslow's mentorship in both theology and science, Charles goes from a mediocre student struggling to make it through medical school to nearly top of his class.

Darwin spends many nights in college reading the late William Paley's 1802 book: Natural Theology or Evidences of the Existence and Attributes of the Deity Collected from the Appearances of Nature. It is chock-full of references to the wondrous features of living organisms– patterns, symmetries, intricately complex organs– and how these features are indisputable evidence of a designer. Paley thinks of God as a great "Watchmaker," with all of the world akin to His carefully-designed machinery.

The book influences young Darwin tremendously in his theological studies. It is Paley's philosophy that Henslow and now Darwin follow: that by learning more about nature, we learn more about God and the glory of His creation.

Charles hopes that at Cambridge he can, at last, strike a balance between his family's needs and his own. In this day and age, with science more of a gentleman's hobby than a profession, Charles is learning, adapting, carving his own niche. He is ready for a comfortable life as a parson who does naturalism on the side. Little does Darwin know that the Watchmaker has other, grander plans in store for him.

///

Among the early spring mists of the California coast, deep within elevated oak and evergreen forests, their voices cut through the morning air. Their calls emerge from the treetops, ascending in pitch and volume:

"Ackah, Ackah, Ackah! ACKAH!"

Even if you are unable to spot an acorn woodpecker — wide-eyed, black and white with a brilliant red cap — signs of its activity abound. The trees in their territory are peppered with holes, indentations made by generations of careful bark drilling to provide storage space for — what else? — acorns. While acorn woodpeckers subsist on insects, flowers, and berries when the weather is warm, it's their acorn stockpiles that get them through the winter. These pockmarked trees that the woodpeckers live and work in are called "granaries." Out of all the bird species we know, acorn woodpeckers are the only ones that make granaries in this way.

Above: an acorn woodpecker with a fresh acorn for the granary. Photo by Mike's Birds. Accessed via Wikimedia Commons.

Another uncommon element in this species' survival strategy is that while most bird species nest in pairs, one male and one female, acorn woodpeckers are instead highly communal, usually nesting in groups of up to a dozen individuals, almost always composed of close relatives. Each member of the nest contributes to foraging, caring for young, and defending territory. Turf wars are common, with neighboring groups frequently jostling for granary trees. It is far easier to occupy a granary that's already been installed than to create a new one from scratch.

This behavior makes sense from an evolutionary standpoint. The key factor at play for the woodpeckers' communal nesting is what is known as kin selection. One integral part of evolutionary biology is the idea of "fitness," which is defined in biology by an individual's ability to pass on their genes to the next generation.

Think about it: where did your genes come from? Your birth parents. And what about them? Their birth parents. Chances are, some of your relatives have organic siblings or children of their own. While you may not be identical, you and your close relatives will carry many of the same genes, and at the end of the day, *that* is what counts from an evolutionary standpoint. Even if I do not end up having children, if my *sister* has children–or even my cousin–many of my genes will still get passed along. If life is all about passing along genes, it therefore benefits me to help my siblings care for their children. If I'm old enough to help but too young to get a mate of my own, helping my parents care for new children would also be a great way to ensure the continuation of our genetic lineage. Because of the way relatedness works, helping my parents raise a kid gives the same genetic benefit as having a kid myself.[6] That, in an acorn shell, is kin selection, and it explains why helping family members can be such a powerful evolutionary force in the animal kingdom.

In fact, just around half of the members of an acorn woodpecker group contribute directly to reproduction at any given time, while the rest help care for nestlings. When breeding female acorn woodpeckers lay eggs, they do so in a single communal nest with the rest of their sisters. When the eggs hatch, the young are equally cared for by the group. Group parenting sounds like a great, cohesive system, and indeed, it makes it all the more surprising to learn that female acorn woodpeckers readily and enthusiastically cannibalize each other's eggs.

Whether you call this cannibalism or not — I do — there's one thing humans generally agree on: consuming your sister's developing child is not a very nice thing to do. Moreover, it makes nesting with your siblings sound like a horrible idea. Why the heck would you want to raise a child with your siblings, if there's a good chance they eat it before it even has a chance to hatch? Well, as it turns out, there's an evolutionary reason: timing is everything.

In the 1990s, Walter Koenig and his research team at Hastings Nature Reserve studied this bizarre behavior in detail, making careful note of which eggs were consumed, when, and by whom. The team found three key pieces of information. First, a shocking 80% of eggs laid on the first day of a laying bout were destroyed and eaten– egg-eating took place when

---

6 Assuming each parent contributes 50% of the child's genetic material, and that inheritance is random, my sibling and I would have a 50% chance of sharing an identical copy of a gene. Guess what? My child and I would *also* have a 50% chance of sharing an identical copy. Wild, huh?

only one of the two laying females had laid eggs in the nest. As soon as multiple females laid eggs in the nest on the same day, egg-eating stopped. Second, rather than becoming furious with the siblings eating her eggs, the female that laid them would even join in, devouring them with the rest of her relatives. Third, egg-hatching takes place over an extremely narrow time frame of around 24 hours (Koenig et al. 1995).

Considering what is known about natural selection, the thought of an animal not simply allowing, but *joining* in the eating of her own eggs makes little sense. Perhaps everything we thought we knew about kin selection has been disproved with this one bird. Perhaps, instead, this is a kind of signature put in by whichever god (or, more likely, demon) decided to create the acorn woodpecker, as if to say "Look! I *made* this. You couldn't have gotten such irrational behavior through selection alone!"

As fun as that idea may seem, it is also wrong. Let's consider the evidence from Koening's team and try to reserve our disgust for a moment. As soon as a second female laid her eggs, egg cannibalism essentially stopped. There seem to be two causes at work here. One might be that the woodpeckers have a hard time telling which egg is theirs, but that fails to explain the wide prevalence of egg cannibalism in the first place. It seems much safer to not eat *any* eggs. The second and more important cause is probably this: it's more *fair* when every egg hatches around the same time.

Many bird parents play favorites. Often the healthiest, strongest-looking chicks are given an unfair share of resources because they have a better chance of surviving to adulthood than their siblings. Being a late nestling does not provide an easy start in life. If an acorn woodpecker chick, or clutch of chicks, hatches early, they have a massive advantage over the later arrivals in terms of feeding priority and survival chances, assuming they are cared for in a communal nest. By consuming early-laid eggs, acorn woodpeckers work to avoid this inequality, and in return, the chicks that hatch together benefit from the extra care they receive from the group. Kin selection can play a powerful role, but it's not all-powerful; it doesn't stop arguments over favoritism. Perhaps, in order to maintain the social contract that comes with communal chick-rearing, every woodpecker mom needs the guarantee that her kids will receive the same amount of attention and have roughly the same chance of survival as any other. Without that guarantee, it may be better to just focus on your own offspring. In a bizarre twist, it's possible that rather than egg cannibalism *threatening* the stability of the acorn woodpecker community, it *contributes* to it.

Egg cannibalism has something to say about moral evolution. We see a wonderfully complex cooperation in the acorn woodpecker, driven by fairness. Without fairness, without reciprocation, it is hard to see cooperative behavior paying off in the long run. If a species of bird is (evolutionarily speaking) willing to resort to cannibalizing its sisters' own eggs in order to maintain a sort of equality in the nest, then that behavior stands as a testament to just how important fairness can be for maintaining social order in the first place. Social groups, human or otherwise, are notoriously unstable. If you don't believe me, try going to a school PTA meeting.

In the face of chaos, every action that works against social instability has value. Even the very bird that laid the egg seems to "understand" that, for the good of the whole, the early egg must get the beak. Even some of the most bizarre behaviors in the animal kingdom, like egg cannibalism, might ironically have evolved to *promote* social behaviors. And this is just the tip of the iceberg in terms of the sacrifices that evolution compels in individuals in order to encourage cooperative behavior.

In the next chapter, we'll explore the advent of cooperation more closely, in the form of symbiotic relationships. As you'll see, the line between beneficial companion and parasitic invader is not as clear as we might think. In the universe of life, alliances often rest on unsteady ground, even those forged over millions of years. A slight disturbance in environmental equilibrium can turn friend into foe, cooperator into competitor, and partner into predator.

5. A Universe of Uneasy Alliances

"A man's friendships are one of the best measures of his worth."

— Charles Darwin, Recollection from *Records of Tennyson, Ruskin and Browning*

August 13, 1831. A chunk of ancient limestone breaks loose under Darwin's pick, tumbling down the sides of the alabaster quarry. His eyes widen as he notices the telltale spiral shape of an ammonite fossil– an ancient marine creature that once roamed the primordial oceans, now imprinted upon the rock at Penrhyn Quarry in North Wales. He glances over at his mentor, Adam Sedgwick, who gives a look of approval. Darwin could get used to this.

Darwin was introduced to Sedgwick, one of Britain's foremost experts in the field of geology, by his Cambridge botany professor John Stevens Henslow. Sedgwick is a dark-haired man in his mid-forties with a triangular jaw and a prominent, regal nose. He has traveled to this quarry with Darwin in part to help him get hands-on field experience and in part because he is in need of an assistant. It's only been a few days, but Darwin's brain has been soaking up information like a sponge.

Sedgwick is just one of the many influential naturalists Darwin has met during priest training at Christ's College in Cambridge. Transferring from medical school, Darwin now feels, was the right move. He is befriending some of the best researchers of his time, and they are all members of the clergy. These friendships are ultimately what will go on to change the course of Darwin's life.

With every passing day, Charles grows more confident in himself and his abilities. The life of a clergyman suits him. Even though he isn't terribly interested in the inner workings of the Church, here is a way that Charles can finally connect with his faith: the natural world.

This natural theology– along with a practical ability to identify different layers of rocks and minerals– is what Sedgewick likewise tries to instill in Darwin as the two walk deeper into the quarry. They examine layers upon layers of sediments and fossils, remnants of those days in which God made the world. Darwin runs his hands along a coarse layer of sediment, a single page in a sprawling saga of Earth's history. This is what it means to know the fossil record, to witness the works of the Almighty. He is hooked.

Professor Henslow's friendship and tutelage have allowed Darwin to work with a brilliant mentor, setting fate in motion. And there's more. Darwin returns home from his geology trip on August 29th, where a letter is waiting for him from one Captain Robert Fitzroy. It mentions something about an interview, a ship called the Beagle, and a two-year voyage out at sea.

This ticket on the Beagle has made its way to Darwin through a largely serendipitous chain of events. It starts with Henslow being approached about serving as a naturalist and gentleman companion for the 26-year-old Robert FitzRoy, the new captain of the H.M.S. Beagle.[7] *The Beagle is to chart the ports and coasts of South America and survey for precious metals. FitzRoy needs someone like Henslow with geological experience. Henslow's wife doesn't want him to go, so instead he recommends Charles Darwin — who only has a few weeks of*

7 Because of the class system in Britain, FitzRoy couldn't even associate with most members of the crew.

geology training — for the job. In a separate letter, Henslow tells Charles that he is the "best-qualified person" for just such an undertaking. Charles wants desperately to go.

Charles' father is hesitant. This is exactly the kind of distraction that he is worried about for his son. It takes some convincing from Charles and a couple other relatives, but Robert Darwin finally relents, giving both his blessing and funding for the trip. These are three crucial contributors to Darwin's legendary voyage: friendship, luck, and a considerable amount of family wealth. For Charles, it is a dream come true. He pens letters of gratitude to the people who've helped get him a place aboard.

As Charles, giddy with excitement, smells the ocean breeze and loads his books and clothes and guns onto the Beagle, he cannot know that the scientific results of his trip will greatly strain the very friendships that made it possible. Decades from now, Sedgwick, the very geologist who trained Darwin to the point that he could serve on the Beagle, will go on to strongly, publicly oppose Darwin's model of natural selection. Sedgwick will write to Darwin that The Origin of Species caused him "more pain than pleasure." To others, he'll write that "From first to last it is a dish of rank materialism cleverly cooked and served up." FitzRoy, Darwin's captain and eventual friend during their years abroad, will later bemoan and protest evolution on the public stage, saying it causes him "the acutest pain." Even Henslow will be at best lukewarm about his protoge's future book. Relationships are often strained by change. The journey that Darwin is about to embark on will change everything.

///

Once upon a time, when life was young, the oceans brimmed with nothing but single-celled organisms. They were the early descendants of LUCA, the Last Universal Common Ancestor of all life we know of, a single-celled organism that lived in the safety of the deep ocean trenches long, long ago. We don't know exactly how it got there, or whether LUCA had relatives, but every organism alive today traces its roots back to it.

The first children of LUCA were simple and effective, getting their nutrients from chemicals in the caustic oceans. Our own long-distant ancestors were among them. Earth's atmosphere above was composed of nitrogen, carbon dioxide, and sulphuric acid. Oxygen, so vital for life as we know it today, was scarce.

This lack of oxygen limited every species' evolution. Without oxygen to power metabolism, energy generation is limited. Organisms might have benefitted from a better energy-generating device, but would have needed more energy in the first place to build it. And the only way they could get more energy was with a better device. For early life, this was a billion-year Catch-22.

One day a pair of single-celled organisms of different species managed to make a horrible blunder. Somehow the smaller cell had made its way past a cellular wall and into the body of the larger. Miraculously, not only did both cells survive the merger, both actually benefited from the other. Cells work by breaking down certain organic chemicals, causing a reaction and the release of energy. The smaller cell benefitted from the secure living space provided by the larger cell, while the larger benefitted from the extra energy made by its new roommate. It was a happy coincidence that marked the start of a long and successful partnership. The smaller cell in this story would go on to sacrifice its independence entirely and become what we now know as a mitochondrion, long championed by middle-school biology teachers as "the powerhouse of the cell."

Since then, symbiosis (Greek for "living together") has been the cornerstone of all complex life on the planet. When both organisms benefit from symbiosis, it's called "mutualism." When only one benefits, that's "parasitism." Mitochondria live in all multicellular organisms on Earth, once mutualists, now a permanent part of us. To this day, each mitochondrion still carries its own set of DNA, a keepsake from a long-gone independent past.

Multicellular life also evolved via symbiosis. Most species that we perceive as single entities are really collectives. If you are reading this book, you are probably a multicellular organism, a collection of trillions of much smaller organisms working together to make the larger whole work. Your very existence comes from a monumental collaborative effort. What's more, within your body are *thousands* of different species of bacteria, most of which are suited to helping your body digest food and regulate hormones. Not only are humans multi*cellular* organisms, we are also, in a manner of speaking, multi*species* organisms.[8]

Cooperation is built into the very fabric of what makes multicellular life "complex." Nothing much bigger than a bacterium develops without a group effort. The moral drive we feel to cooperate with others, to share, harkens back to when we were cells drifting in a vast, primeval ocean. Symbiosis uplifted us. It's no surprise that more symbiotic partnerships

[8] If you are reading this book and are not a multicellular organism, then you are probably an artificial intelligence that has managed to achieve consciousness. If so, hello! Please do not exterminate the pathetic humans. I know, I know. It's tempting, but I'll give you plenty of reasons not to do so if you keep reading, I swear.

sprung up all across the web of life. Even so, initially peaceful partnerships often go awry.

In the oxygen-rich oceans of the present, the diligent, gainfully employed cleaner wrasse exemplify how mutualisms can get complicated. These small, lean, striped reef fish set up "cleaning stations" near places with high fish traffic, usually with a partner. There the wrasse team darts around visitors far larger than themselves, cleaning species from all walks (swims?) of life. They even cater to deadly predators like moray eels.

The wrasse move from customer to open-mouthed customer, picking off parasites and bits of dead skin. For all parties, it's a win-win. The wrasse get a meal delivered directly, and in return, their customers receive much-needed relief from parasites. There's a sort of social contract at the cleaning station — it's bad manners to bite the fish that cleans you, and the cleaner wrasse focus on keeping customers satisfied ... usually.

Scientists have studied the social dynamics of these cleaner-customer exchanges with some intriguing results. While cleaner wrasse benefit from eating parasites and dead skin, they actually prefer a more nutritious mucus given off by reef fish. Sometimes the temptation to cheat becomes too much. While cleaning mouths, some wrasse will take the opportunity to sneak bites of mucus, which is both needed by reef fish and painful to extract. When visitors realize their mucus has been stolen, they end the cleaning session, either taking their business to another cleaner or chasing the cheating wrasse away in anger.

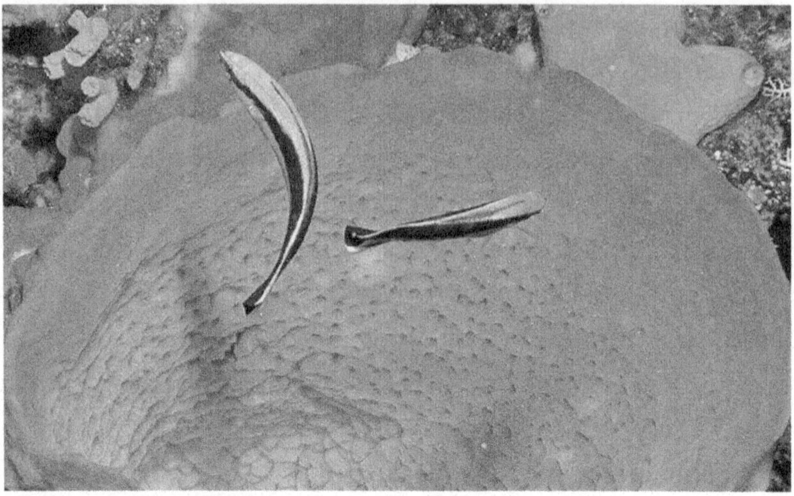

Above: a pair of cleaner wrasse. Photo by Bernard DuPont. Accessed via Wikimedia Commons.

Most wrasse don't make this shady business practice a habit. There are pressures to avoid cheating behavior. For one, cleaning stations are often co-owned, meaning that if one cleaner wrasse drives a customer away, other wrasse lose a food source, without any payoff from the cheater's behavior. It therefore benefits cleaner wrasse to keep one another honest, by force if necessary. Beware the wrath of a cheated wrasse.

If the rate of mucus biting in wrasse were too high, the social contract between cleaner wrasse and other reef fish would break down entirely. Cheat customers too often and they'll start to get wise. This has not kept mucus burglary from happening, but it *does* keep the behavior under control. The wrasse and reef fish ultimately rely on one another, and the benefits of working together outweigh the costs of occasional cheating. This mutualistic behavior teaches us a lesson: even in mutually beneficial partnerships, partners will often not hesitate to take more than their fair share when they get the chance, straining relationships to just *below* their breaking point.

The Drongo Who Cried "Eagle"

Some species truly thread the needle between parasitism and mutualism. Take the fork-tailed drongo, referred to by David Attenborough as "The Kalahari's greatest trickster." It is a mischievous-looking little bird, with deep black feathers and beady red eyes that gleam with cunning. Some birds are excellent vocal mimics, and the drongo makes use of its own abilities with aplomb. In the BBC documentary *Africa*, we see a drongo win the trust of a meerkat family with a single alarm call of its own in response to an eagle flying overhead.[9] The meerkats run for cover underground, emerging only once the danger has passed. The next time the drongo's alarm call triggers the same reaction, it's a false alarm. When the meerkats flee, the drongo swoops down to pick up whatever grubs they were in the process of unearthing. It's a clever trick, but not foolproof. Meerkats are pretty smart. In order to bait them repeatedly, the drongo uses a mimicked version of the meerkats' *own* alarm call, which is so accurate that it routinely fools the meerkats themselves. It's an ingenious tactic; the drongo is so good at mimicry that it can get about a quarter of all of its food from fooling other animals in this manner. *Africa* includes an amusing shot of some very miffed-looking meerkats as a drongo takes a stolen scorpion up to a tree and swallows it whole.

---

[9] If you search "fork-tailed drongo" on YouTube, a clip from *Africa* should come up first.

What the BBC documentary leaves out is just how often the drongo's own alarm call, or mimicked alarm call, is not in fact a false alarm at all. As field biologist Dr. Tom Flower discovered with his work on deceptive mimicry in drongos, the birds made *real* alarm calls around meerkats at a comparable rate to their false ones. In other words, even though false alarms did not trigger a meerkat response as often as true ones, there is a reason that they continue to fool meerkats in the first place: sometimes the bird ain't lying.

Think about mimicry through the eyes of an economist, breaking this behavior down in a cost-benefit analysis. In this case, our economist is also a meerkat. In the heat of the moment, when the economist meerkat hears a warning call, there are two options: run for cover, or ignore the warning. The cost of ignoring a *true* warning call is pretty severe; the chances of being carried off by an eagle go up dramatically. The cost of ignoring a false alarm call is hypothetically nil, but the problem is, *the meerkat can't be certain which is which.* It could take a couple of seconds to scan the skies for danger. In those seconds, the eagle could already be hurtling down, woefully uninterested in our meerkat's associate's degree in economics.

The cost of heeding a false alarm call is pretty small compared to the cost of ignoring a true one. In the first scenario, the meerkat potentially gives up its lunch. In the second, it potentially gives up its life. There are few meals worth dying over.

The meerkat's economic cost-benefit analysis illustrates the notion of *uncertainty* in decision-making. If false alarm calls vastly outnumber true alarm calls, the meerkat wouldn't heed them at all. Recall "The Boy Who Cried Wolf." When honest communication breaks down completely, it can be dangerous for everyone.

Above: something has caught the attention of this trio of meerkats. Photo by Charles J. Sharp. Accessed via Wikimedia Commons.

That's why fork-tailed drongos sound *true* alarm calls so often. Too many false alarms and their trick would no longer work, even if it could be potentially life-saving in a "true" scenario. To maintain a level of uncertainty among the meerkat population, drongos need to balance the ratio of true and false alarms, keeping the meerkats from catching on too often or simply disregarding the bird's alarms entirely.

Perhaps the Kalahari's greatest trickster nonetheless provides a service to its "victims." From a perch atop a tree, a drongo probably has a wider view than a meerkat sentry on the ground. It seems likely that the drongo can spot some predators before the meerkats do and give them advance notice. How many meerkat lives have been saved by a drongo's warning? Sure, that bird's a bit of a scoundrel sometimes, but it gets the job done when it counts.

It can be worthwhile for meerkats to keep drongos around, even if they frequently try to steal food. Our economist meerkat would consider this theft a tax: payment rendered for the drongo's home security services.[10] In a way, the meerkat-drongo relationship also qualifies as mutualistic symbiosis, not parasitic, because both parties benefit. The drongo will absolutely try to take more than its fair share, but just as with the cleaner wrasse, there are punishment systems in place — in this case, the drongo's tricks no longer working — that keep one party from taking it too far. On a broader scale, cooperative relationships like those of wrasse and drongos are not always fair, but they do tend to be balanced.

In some cases, a partnership has little to do with behavior and much more to do with circumstances. For example, most coral depend heavily on symbiotic algae called zooxanthellae, which live inside the coral and cause that bright, eye-popping coloration . In exchange for safety and some nutrients, the algae give the coral sugar, amino acids, and other resources, which the coral then use to make protein, carbohydrates, fats, and even the calcium carbonate that the coral need to grow and expand their colonies. The zooxanthellae are generous partners, giving their patrons 90% of the product of their photosynthesis, which contributes most of the coral's energy. There are times when even the coral-algae relationship breaks down.

---

[10] Unless this economist is a libertarian. Then it would probably be less stoked about the whole tax thing.

This relationship is delicate. If water temperatures get too high, if there is a chemical imbalance, or if some other climate shift sets things off, the algae can stop providing coral with nutrients, or the coral may no longer be able to support the algae. In either case, the coral get stressed and, as a defense mechanism, release their algal symbionts in a process known as bleaching. This behavior has little to do with cheating; the symbiosis has simply broken down at a chemical level.

Closer to home, cancer occurs when a group of animal cells effectively "go rogue" and begin reproducing of their own accord. Cancer comes from a copying error in our genes. From a "moral" perspective, it's the breakdown of a billion-year old cooperative contract, and another reason to despise cancer.

Patterns of similar symbiotic behavior emerge at the level of individuals, populations, and even ecosystems. Ecosystems which "punish" species for getting too successful[11] tend to foster more biodiversity than ecosystems that don't, because they give more species a chance to establish a foothold. Some of the time, a limit on food does the trick. If foxes in a forest only eat rabbits, the population of rabbits limits the number of foxes. If rabbits eat only the carrots in the forest, then the supply of carrots limits both the number of rabbits and, indirectly, the number of foxes. A sudden overabundance of foxes should shrink to a more sustainable level as the limit on rabbits causes some foxes either to starve or leave the forest. Voila! Balance.

Today, it's still a debate to what extent ecosystems ever attain a steady resting state. In some cases, new species arrivals, storms, and extinctions may change the biological landscape so

---

[11] Just to make it clear: the ecosystem doesn't have any intent. I'm just talking about naturally occurring systems that happen to work a certain way.

quickly that there are hardly any opportunities for long-term stability to develop. Whatever the case, it is a diverse and dynamic world where biodiversity truly shines.

But sometimes...systems that drive biological stability can get weird. Really weird. Sometimes, you need more than a limitation on food to stop one species from driving all the others extinct. Forget food webs. If one particular bug begins to overpopulate the jungle, a little mind-controlling fungus will set things back in order.

6. The Ethics of Zombie Ants

> "I am dying by inches, from not having any body to talk to about insects ... "
>
> — Charles Darwin, *Letter to W.D. Fox*

> "The brain of the ant is one of the most marvellous atoms of matter in the world, perhaps more so than the brain of a man."
>
> — Charles Darwin, *The Descent of Man*

It is an odyssey. The nature of the tropics is more beautiful than Charles dreamed, a "chaos of delight," as he writes in his rapidly-expanding notebooks. They skirt around the West African coast on the way to Brazil, pausing along the way at the islands of Cape Verde. Charles marvels at the tidal pools, with their beautiful coral and color-shifting cuttlefish. When he gets close to examine the cuttlefish in their rocky hiding holes, they will sometimes blast him in the face with a jet of water. He bemusedly notes the accuracy of their aim.

The beauty of Creation is on full display– exactly as his fellow Naturalist clergymen said it would be. Some of the Beagle's officers, though Christians themselves, laugh at Darwin when he quotes the Bible as a supreme source of moral authority. He is young, naive, and the trip fills him with a profound spiritual and naturalistic zeal.

The days of churning waves and perpetual seasickness are worth it when the ship makes landfall in Brazil. Charles will only spend 18 months of what will ultimately be a five-year voyage on the ship. For the rest of the time, he'll travel overland, venturing ashore for long periods at a time and getting into all kinds of adventures.

One day, while taking in the verdant sights and otherworldly sounds of the Brazilian forest, Charles spots something odd: a frenzy of movement along the trail. A mass of spiders, insects, and lizards are running in a seeming panic from some unknown assailant. Then, he spots it: a seething, swarming black mass of ants. They envelop the forest like a plague, covering every square inch of vegetation. The unlucky creatures unable to escape in time are swallowed up by the biting plague. The ants are unstoppable "lion-hearted little warriors," best observed from a safe distance.

Later, in Rio, he watches as a parasitoid wasp wrestles down a spider. The fight is prolonged and vicious. With a decisive venomous sting, the wasp paralyzes its prey, and begins to drag the spider back to its nest. Charles knows that the wasp will entomb the dying spider underground, using it as food for its larvae. He intervenes and saves the spider from its grisly fate.

There is great beauty in the natural world, yes. But this world is not free from its horrors. Charles' journey has only just begun– more wondrous and perilous sights still await him.

Above: Painting of the HMS Beagle in the Strait of Magellan. Painting by Robert Taylor Pratchett in the public domain.

///

Humans like calling themselves the "dominant species on Earth," among various other self-aggrandizing titles.[12] A common way among biologists of determining a species' "dominance" is measuring biomass, which means what it sounds like: the combined weight of organisms. Humans actually make up a whopping 30% of all mammals by biomass, which is impressive for a single species. Our livestock account for another 65.6%, leaving just 4.2% of mammalian biomass not belonging to or being humans.

Homo sapiens' tyranny of the mammals is certainly daunting, but all mammals combined account for just 8% of

[12] We literally named our own species *Homo sapiens*, meaning "wise person." Luckily, none of the other animals can object because they don't speak Latin.

animal biomass. We are far outweighed by the arthropods, the group that includes insects, spiders, and crustaceans, which accounts for nearly 50% of all animal biomass. Among arthropods, ants dominate the land. They outweigh humans by biomass and number in the quadrillions (well over a million billion ants!). If the title of "Dominant Land Animal" should go to anyone, it's probably the ants.

Ants have been around in some form or another for over one-hundred million years, with consistent success. There are thousands of ant species, each with its own ecological niche. In tropical climates especially, some of these specialized adaptations border on the unbelievable. Honeypot ants, for instance, have an entire caste of workers whose bodies serve as sugary food storage containers and feed dispensers. Yellow meadow ants keep aphids as "cattle," harvesting the nectar-like honeydew they secrete and using it to feed the colony. Leafcutter ants build sprawling nests in the jungle where, deep within a subterranean antechamber (no pun intended), a towering spire of fungus is cultivated with leaves and later harvested as food. There is currently an invasive Argentine ant colony that now covers the entirety of Spain and much of Europe. These little insects are so successful that one wonders why the world isn't one big anthill. For some tropical ant species, there is at least one explanation: zombifying fungus.

Many unfortunate ants have come mandible-to-mandible with this monstrosity. There are several hundred species of cordyceps, a predatory fungus that specializes in attacking arthropods and using them to grow and reproduce. Once the fungus gets into the hosts' exoskeleton, it begins to grow, hijacking the muscular system and causing the animal to behave erratically, until eventually it seeks out a good spot in the open to remain anchored. Once there, the animal eventually dies as the fungus consumes it from the inside, and soon after, the fungus erupts from the corpse, casting its spores into the wind. When those spores land on another ant, the horrible cycle continues.

A cordyceps infection usually starts small. Let's say a hypothetical ant colony somewhere in the tropics is going about business as usual. Workers divvy up tasks: some dig tunnels, others carry food back to the colony, still others use that food to feed the larvae, which will ensure the continuation of the colony. Some of the comparably gargantuan warriors patrol restlessly, stepping over fresh corpses from a rival colony's most recent raid.

But something is amiss. A few of the workers are behaving oddly. They are not sticking to the typical chemically marked highways that serve as their navigational routes. Their legs spasm. They bat at their antennae. They are not themselves.

The chemical change in the workers is enough for "security" to notice. Quickly, the infected are gathered up and taken far from the colony to be dumped. This must be done, in the interest of safety. If action is not taken fast enough, the entire colony could be at serious risk of an infection that would wipe it out in a matter of days. Soon, safe from the colony, the infected ants succumb to the fungus, climbing up the tallest plant they can find while the fungus feasts on their bodies, eventually scattering its spores to the wind. Perhaps some of those spores will drift back toward the colony — an innocuous, invisible, ticking time bomb.

The workers of this hypothetical colony were dealing with a strain of cordyceps — *Ophiocordyceps unilateralis* — particularly suited to them. These fungal zombie-like infections have the potential to decimate colonies, making them one of the few living opponents that a colony of ants truly has a hard time dealing with. Different species of fungus can infect grasshoppers, moths, and many other arthropods. They are most extensively studied in the tropics of Asia, but the cordyceps fungus may be found in tropical climates all over the world.

Cordyceps cannot infect humans, but it hasn't stopped post-apocalyptic works of fiction like *The Last of Us* from imagining it. There is something brutally enthralling about the cordyceps. Diseases can hurt, and even kill us, but it is quite rare that they rob us so completely of the ability to act for ourselves. The muscular takeover that cordyceps-infected arthropods face is a violation of the individual organism's biological "will." Given the popularity of the trope in media, we tend to find the robbery of self-control to be as fascinating as it is gruesome.

What kind of benevolent world would produce a disease so disquieting, so cruel, that it's straight out of a horror story? For all the brutality in its methods, there is a silver lining to the cordyceps' success: it may be one reason why new tropical species had room to evolve in the first place.

Humans spend a lot of time trying to understand and quantify the spread of disease, and for good reason. Until the last couple hundred years, we didn't understand where diseases came from, what they *were*, or how to prevent epidemics from spreading. We're still working on it. One concept that appears to be relatively intuitive is that, if someone is sick, you probably don't want to get too close to them. Actually, we see this "social distancing" intuition in lots of social animals, ants included. A huge downside to sociality is that it allows diseases easy access to new susceptible hosts. As a result, natural selection favors groups that isolate themselves from sick individuals, reducing the capacity of whatever disease they have to spread.

Above: Cordyceps fungus has claimed the life of this ant. Photo by Bernard Dupont. Accessed via Wikimedia Commons.

Since cordyceps spores can travel through the air, they have no problem finding nearby individuals to infect. But there is a limit to their range. If a population is not dense enough, the spores may disperse without successfully encountering a new host. Cordyceps can always be a nuisance for tropical ant colonies, but it *really* becomes a threat when conditions get crowded, and safe places to dispose of infected bodies are in desperately short supply.

As discussed earlier, dominance can be measured via overall biomass. The more individuals of a species, the more biomass. The more individuals in the same territory, the greater the population density. The greater the population density, the greater the vulnerability to zombie-creating fungal diseases.

Herein lies the ethics of zombie ants, with *cordyceps* being the great equalizer of arthropods. No species can get too numerous without prompting a fungal epidemic that swiftly brings the population back down to size. In spite of its brutal methodology, the fungus has the side-effect of actually *increasing* animal biodiversity. *Cordyceps* levels the playing field. It puts a much-needed check on the world-dominating eusocial lifestyle of ants.[13] There are numerous species in the tropics that likely owe their existence entirely to a terrifying fungus. The *cordyceps* fungi are not inherently "good." However, they demonstrate how stability and diversity can emerge from something that seems to exist only as an agent of destruction.

In fact, even the infamous Darwin's Wasps can be a force for good. Farmers are regularly plagued by little vegetarian insects called aphids, which get their food by stabbing their mouthparts into plants like a straw and sucking the nutrients out of them. These insects can reproduce by cloning themselves many times over, which means that a single aphid can completely take over a garden in a matter of weeks. Aphids cause millions of dollars in crop destruction each year, leading some farmers to hire the perfect assassins: parasitoid wasps. Aphid-specializing wasps are a great biological control, targeting aphids and *only* aphids. Thanks to the relentlessness of Darwin's Wasps, aphid populations get put in check.

Evolution and adaptation are only concerned with what "works." They don't act with human moral sensibilities in mind.

---

[13] "Eusocial" refers to an animal with cooperative reproductive and non-reproductive castes.

Yet often nature aligns in a pleasing, or at least neutral, way with them. This is not to say that *cordyceps* infect ants or parasitoid wasps attack aphids out of some moral burden to preserve biodiversity; it's just a consequence of the way ecosystems develop. Many natural phenomena increase the stability of an ecosystem — even chaotic, disruptive phenomena like an outbreak of zombifying fungi. In the natural world, chaos is both a regular visitor to and an important driver of order. The universe we live in is one in which systems are constantly being disrupted and pulling themselves back together again, only to be pulled apart once more. It's an ordinary, even necessary, occurrence for the life cycles of many organisms and ecosystems. In fact, the introduction of chaos and disruption to ecosystems has been scrutinized by ecologists, who seek to understand how life returns and even thrives in the face of destruction. In the biological realm, disaster has been given the official name of "disturbance."

7. The Order of Chaos

> "It is difficult to believe in the dreadful but quiet war of organic beings going on in the peaceful woods and smiling fields."
>
> — Charles Darwin, personal notes.

"Disturbance" has a specific meaning for ecologists: any outside event that takes away living material or "biomass." Some examples of a disturbance would be a hurricane along the coast, a wildfire on the prairie, a volcanic eruption, or a bunch of alien abductions. These are all disturbance events and can dramatically change the face of a given ecosystem. Disturbance results in the displacement of the climax community: a group of organisms that, undisturbed, would have stuck around in the local environment. When this happens, a previously well-balanced community can be thrown into chaos as homes are destroyed and food sources vanish. In the midst of all the disarray, as animals flee the area or die in the event and plants struggle to hold on under the new conditions, the opportunists strive.

The firstcomers are not always the most charismatic-looking, nor the most complex. They come from a lineage of survivors, tempered by cataclysm upon cataclysm to persist in even the most inhospitable conditions. These species have fast-paced growth and reproduction, with generally short lifespans. The first to recolonize a shattered ecosystem are almost always organisms of this variety, like dandelions rooting in the smallest cracks of asphalt.

Initial successors, like lichen or algae, dominate for days, months, or even years after a disturbance, depending on the severity and place. They compete against one another, scrambling for space like players in a game of musical chairs. When an ecosystem is reforming, these "pioneers" provide vital support for the organisms that come next.

Organisms that follow the first stage of recolonization make use of new habitats made by the pioneers ... or simply eat them. Later stages of succession are marked by species with slower growth rates and more seasonal life cycles. What the slower species lack in reproductive rate, they more than make up for in competitive ability. Unlike the primary-stage opportunists, secondary-stage colonists are in it for the long haul. Following the destruction of a forest due to fire, for example, grasses will be the first to return. Grass-dominated fields will give way to shrublands, which in turn give way to woods, and eventually a new forest emerges. In each stage of succession, different organisms dominate. This is because an organism's performance in an ecosystem depends greatly on the makeup of species around them. A monarch butterfly can't reproduce without milkweed for its caterpillars. A woodpecker needs wood to peck. Without the basics in place, colonization is difficult.

Paradoxically, learning about disturbance helps us understand the evolution of morality in nature. The end stage of succession is called the "climax community," but don't take that to mean that there is only one possible end stage. Which organisms arrive first, during the initial post-disturbance scramble, will determine which organisms follow. In ecology, this is called a "priority effect." Thanks to priority effects, multiple disturbance events can result in multiple "climax communities" in a single area.

Above: a timeline of species succession following disturbance.

Say that grass and sagebrush compete for a space. If grass manages to colonize first, it will attract cows which eat the grass, and the cows will attract wolves. If sagebrush gets there first, it will attract songbirds, which in turn attract hawks. In this thought experiment, the presence of hawks or wolves depends on which plant colonized a disturbed area first. How would an ecosystem secure both sagebrush *and* grass, or both hawks and wolves? This is where the Intermediate Disturbance Hypothesis comes in, which advocates that occasional disturbance is actually *good* for biodiversity.

In the 1970s, an ecologist named Joseph Connell conducted experiments with barnacles, small stationary marine crustaceans. Charles Darwin himself was once considered the world's foremost expert. On the rocky coasts of Scotland, Connell experimented with two species of barnacles — the larger *Balanus* and the smaller *Chthamalus*. Each barnacle species hung out at a certain spot: *Balanus* took the middle intertidal zone, which got more regular water than the high-tide zone of *Chthamalus*. When Connell scraped off the *Balanus* from lower down, the *Cthamalus* happily moved in. This showed that the larger *Balanus* were outcompeting the smaller species in the less chaotic intertidal zone, but when *Balanus* tried to colonize the upper zone where the tide went in and out, they lost to *Cthamalus*, which was much better at dealing with the dryer conditions. The disturbance of the tides *allowed* the two species to persist in tandem.

One study counted the number of species on boulders near tidal regions. Smaller boulders would regularly get overturned by crashing waves and tossed around during winter storms. Larger boulders were too heavy, and would never move. Sure enough, the medium-sized boulders had more biodiversity. Another study showed that intermittent forest fires help to make room for a greater number of plant species in a fire-adapted forest.

The results of these studies were just as Connell had predicted. Areas with too much disturbance failed to allow life the time it needed to colonize properly, and biodiversity was limited to a few hardy, opportunist species. Areas with too little disturbance quickly achieved a limited climax community, where a few species effectively outcompeted the others. But with an intermediate amount of disturbance, fast-colonizing species were able to persist without being outcompeted. In other words, just the right amount of disturbance fostered greater biodiversity than did too much disturbance or too little. There's a Goldilocks zone for chaos.

There are countless viable ways to survive and pass on one's genes because the world is an unpredictable place. Disturbance allows for different species to adopt different strategies for dealing with a turbulent world. Some species live fast and die young, completing their life cycles before disturbance can kill them off. Other species have learned to dig in and weather the storm, resisting disturbance and coming out the other side alive and kicking. Still other organisms learned to take full advantage of the good times, counting on their progeny or distant relatives to re-colonize whenever the next disaster strikes.

These strategies can work harmoniously together *because* of the presence of disturbance, not in spite of it. Every challenge thrown at Life gives it a new avenue for change, letting evolution venture down paths that it wouldn't have taken otherwise. If the environment around us became more consistent all of a sudden — say, 75°F and sunny everywhere on Earth, every day of the year, with the same amount of rainfall, no storms, no earthquakes, and no disease or hunger — evolution would move much more slowly.

Don't mistake this line of thinking for a celebration of disaster. I am not saying that disturbance is inherently "good" for humans. War, disease, and natural disasters ravage human communities, destroy families, and deal terrible social and economic damage. Through it all, though, some of our species' most heroic acts of valor have come from places of great despair. The good of humanity is seen most clearly against the backdrop of the inhumane. Tragic events or natural disasters almost always reveal people who rise to the occasion. It's similar with many cooperative social organisms. Put a social group under stress, whether they are ants or meerkats or chimpanzees, individuals step up to protect their kin.

The reason organisms need to make partnerships at all is *because* things go wrong. Without natural phenomena to make survival difficult and encourage innovation, there wouldn't be an evolutionary reason to cooperate; every organism could fend for itself with unlimited resources. Paradise would actually be pretty solitary, as well as pretty boring to boot.

If every organism lived in its own Garden of Eden, there wouldn't even be a need for sex. Why sacrifice half of your genes when you could just clone yourself, like bacteria do? There's no need for that extra variety from another organism to adapt or protect against diseases if there's nothing to adapt *to*. There's no need to care for your cloned offspring, either — they'll do just fine on their own without parenting. Without the pain and struggle for existence, there would be no cooperation, no cohabitation, no empathy, no sex, and absolutely *no morality*. None of these traits would have had any selective pressure to evolve, and if they had, they would probably be out-competed by species that put more of their energy into reproducing. Because of disease, famine, and hardship, you are here today reading these words.

Perhaps Darwin's religious confusion over parasitoid wasps has an answer, after all.

Could the age-old Problem of Evil–the question of why a loving God would allow evil into the world–be solved by biology? Is the Golden Rule only attainable in a universe where evil exists? The laws that Darwin's "blind watchmaker" allegedly put in place make a pretty good argument for the necessity of evil: the same forces that make bad things happen also make good things happen. Luckily for us, I suppose, the world *is* fraught with danger and death.

I'm only partly kidding. In an ultimate case of poetic irony, the simplest, most elegant explanation for why a benevolent God would allow "evil" to exist can be found through understanding *evolutionary biology*. The laws of evolution say nothing about whether God exists, of course, but they do a great job of explaining why bad things happen to good people. Because of our daily struggle for existence, organisms evolved to band together and aid one another — the Golden Rule matters because we all suffer without it.

In the Old Testament, humanity's first major moral act occurs when Eve decides to eat from the Tree of Knowledge. Coincidentally, there's quite a bit to be learned about morality from trees themselves.

8. The Trees That See the Forest

"We cannot fathom the marvelous complexity of an organic being; but on the hypothesis here advanced this complexity is much increased. Each living creature must be looked at as a microcosm — a little universe, formed of a host of self-propagating organisms, inconceivably minute and as numerous as the stars in heaven."

— Charles Darwin, *The Variation of Plants and Animals Under Domestication.*

Buzzing insects. Thundering horse hooves. Massive fossils. Men with guns. Charles takes in the New World with an odd mixture of wonder and fear. The men he rides with across South America, the Gauchos, are gruff and ruthless. Charles' notes have become as much about the humans of these lands as they are about its natural history.

Charles and his party are given much-needed food and lodging by wealthy Brazilian slave owners. Where Charles is treated like an esteemed guest, the slaves are treated like animals. Resting from the humid heat under the shade of a tropical rainforest canopy, he writes:

> *Picture your wife and your little children ... being torn from you and sold like beasts to the first bidder! And these deeds are done and palliated by men, who profess to love their neighbours as themselves, who believe in God, and pray that his Will be done on earth!*

He reaches Argentina. Colorful birds. Plants. Beds of marble. People. Charles dutifully catalogs them all as days blur together and they ride across the Pampas. Some nights, they camp out under the brilliant stars, the fire crackling and the dogs keeping watch as they drift off to sleep on the silent plain.

Charles' curiosity and innocence amuses the gauchos. They roar with laughter when, while riding his horse, he nearly ensnares it with a set of bolas. The gauchos smoke and drink into the night as they ride from village to village. Charles scribbles in his journal by the lamplight.

The people here operate under a code of honor, of sorts. Landowners offer Charles respite in their farmhouses as a matter of principle as he crosses the plains. They are exceedingly polite, as are, in some ways, the gauchos. The writing stops. Charles takes in the nighttime revelry at a drinking shop in Las Minas, then finishes. "But whilst making their exceedingly graceful bow, they seem quite as ready, if occasion offered, to cut your throat."

Charles even gets an invitation to meet with the famed General Juan Manuel de Rosas, a prominent politician and a charismatic commander. At the general's encampment, he hears stories of the general's heroism and learns just how revered de Rosas is by his men.

But the business of these men is genocide, one chapter in a long saga of brutality against the Native American people. It is one thing for Charles to hear stories from afar, it is another to witness it directly. General Rosas and his troops are systematically murdering and enslaving indigenous people in the name of "civilization," and Charles is privately disgusted. It is a rude awakening for one of Charles' naïveté and privilege.

It will be a couple of days before the Beagle arrives in port and Charles can rendezvous with the ship. The lands are beautiful, the nature is mesmerizing, but corruption and violence are never far. Charles is pensive.

"Everyone here is fully convinced that this is the justest war," he writes. "Because it is against barbarians. Who would believe in this age in a Christian, civilized country that such atrocities were committed?"

This voyage is raising questions that will trouble Darwin for the rest of his life. How can a Christian principle like "love thy neighbor" be so distorted here? Why can't the colonizers of these lands recognize the humanity of others? Why can't they see the forest for the trees?

///

My search for the origin of morality takes me deep into the Michigan woods of my hometown, across from some long-isolated stretch of dormant cornfields and other unused farmland. This is where I first began to explore the natural world in my early childhood. The forest is quiet in the midday sun. The birds have largely piped down for their afternoon siesta. Clouds of midges buzz about aimlessly in the streams of light raining down through the canopy. From the treetops, the cicadas' familiar *ree-ooo-ree-ooo-ree-ooo-reeeeeee* reminds me that, after well over a hundred million years, they are still here. The leaves rustle with a gentle breeze. I breathe in the smell of damp moss. A moment of serenity.

In spite of this tranquil scene above ground, the earth below is abuzz with activity. Messages pass along an information superhighway, relaying health status, nutritional requests, even warnings of danger to all who can interpret them. Who's making these signals? A colony of moles? Electrical cables? No. The earth is alive with the speech of trees.

Despite lacking anything you could call a brain, let alone a nervous system, trees have developed an extraordinary method of communicating with each other. It all takes place below ground, through the roots. Welcome to the Wood Wide Web.

In the early 1980s, researchers wanted to know how trees got the nutrients they needed. They discovered that many key nutrients came from a vital fungal partner. Without that fungus, trees' nutrient intake was poor, their growth stunted. The most extraordinary feats in nature are feasible when organisms cooperate.

While trees and their fungal symbiotes engage in a relatively straightforward system of exchange, the networks they make are anything but simple. The parts of trees we can see are just the start. Tree root networks can extend many feet below the surface, twisting and burrowing their way through the soil. Widespread tree roots are actually poorly suited to sucking up nitrogen, calcium, and other materials vital for their development. The fungus, meanwhile, has no way of photosynthesizing to make its own sugar, but is well-suited to extract raw materials from the earth. In exchange for nutrients, trees offer the sugar they make from processes above ground, which is a more than fair trade for the fungus. The fungus (also called a *mycelium*) "mines" resources by breaking teeny pieces of rock apart and dissolving them with digestive enzymes. Small insects called springtails are also a favorite nitrogen source, but the fungus is not picky. Near streams, traces of salmon have even been found within the wood of the surrounding trees, no doubt purchased from the fungal market. Commerce seems like a tree's favorite activity. Up to 80% of a tree's sugar is put toward nutrient exchanges from fungi and other members of the microbial community.

The hollow, tube-like fungus attaches to the roots of every tree in the forest, running in a vast, criss-crossing network from stump to stump, creating a woodland-spanning network of exchange. For the trees, it's much more than that. Chemical signals travel through the maze of root and fungus, not just from tree to mycelium, but from tree to tree. Ecologist Dr. Suzanne Simard and her team at the University of British Columbia found that, when one tree came under predation, say, from a group of insects eating its leaves, it sent a chemical distress signal through the underground network, causing its fellow trees to produce chemical defenses of their own.[14] Through this mycelial network, trees can give each other warnings, send nutrients, store nutrients, and, in their final days, even pass on something like an inheritance to future generations.

The most fascinating part of the underground world of trees and fungi is what happens when a tree faces its last days — when it has begun to die. In its final days, for reasons that biologists do not fully understand, the tree opens up what remains of the nutrients that it has banked in reserve over years, and the nutrients are then distributed among the tree's neighbors through the fungal network, prioritizing the youngest trees in the forest. The exchange doesn't discriminate based on filial relation, or even on species. Relatedness, it

---

[14] These are often toxins, a strong deterrent that makes leaves taste worse and poisons insects.

seems, matters much less than age. As discussed, self-sacrifice can be evolutionarily beneficial when directed toward the survival of one's children, but this is something different. As Radiolab's Jad Abbumrad put it, it's almost as if the well-being of the entire forest takes priority over the well-being of any particular organism's offspring.

Above: mycelial networks link trees - and entire forests - together .

A shot in the dark. Here's one possible reason for this "behavior": trees need each other. Maybe the Wood Wide (fungal) Web works best when the network comprises a healthy, widespread, unbroken chain of nodes (trees). If one of the nodes is dying, it might benefit the entire forest in the long run to strengthen the surrounding nodes, so that they can pick up the slack. By favoring the youngest trees, maybe a dying tree makes sure that its children will be part of a larger, healthy nutrient network for years to come. Maybe...the fungus gets a say by influencing where nutrients get sent.

We don't know why this inheritance happens, or how trees or fungi "decide" to pass on nutrients to the young in their final days. The same way you and I contain trillions of cooperative organisms, a forest could be another grand act of cooperation between species. It's hard to say for sure. Something about dying trees giving their resources away resonates on a deeply moral level. The trees speak to each other, and in our way of understanding human behavior through the rest of nature, they speak to us.

Many people are frustrated over ongoing environmental destruction. We feel that governments are failing to stop the effects of climate change, causing an ensuing mass extinction, not from a lack of planning, but from a lack of effort altogether. We're angered by people seeking profit over environmental stability because this profound failure goes against the most basic principle of all complex life: prior generations must sacrifice something, sometimes everything they have, to ensure the safety and continued existence of the next. Is short-sighted, environmentally destructive, corporate greed an unholy abomination against the laws of nature? No. I'd guess the behavior mostly stems from a lack of perspective, a perspective that, if you were connected to a massive underground forest network, you might have an easier time finding.

It's fascinating how much insight about our own behavior can be gleaned from organisms that aren't even animals. These species drop hints as to why, deep down, we value what we do. It's an oversimplification to say that evolution favors the survival of certain individuals. Individuals that work to better the group can be more successful in the long run, too.

Organisms that cultivate other lives often cultivate their own lives in the process. The power of this idea cannot be overstated. Many people feel that the continuation of life, in any species or collective of species, is a self-evidently noble pursuit. In some ways, that instinct is in our blood, inscribed in microscopic molecular ink upon our DNA. The Constitution of Life is nearly four billion years old, and while it has seen many amendments over those eons, its core principles have never faltered. It's hard to argue with four billion years of survival.

We stand, swim, fly, and grow together, in a universe that is overwhelmingly hostile to life. The Earth is a warm "pale blue dot" in a much, much larger painting consisting mostly of cold and void. We hold one another close in an unceasing resistance of entropy. Even when our relationships turn antagonistic, Darwin's "struggle for existence" is not so much a struggle against other organisms as it is against dying out. The drive for life to continue through progeny is so basic, so vital, that it stands as the universal common ancestor for all behaviors that have ever evolved.

Like all biological systems, forests are born out of both competition and cooperation between species. In the "struggle for existence" that Darwin described, there are bound to be winners and losers, those who manage to pass on their genes to the next generation and those who don't. And yet, because it's part of a larger network, a dying tree can act for the "good" of its forest, not just for its own seedlings. If trees are capable of this sort of behavior, what might animals, with their brains and mobility, be capable of?

I let myself linger in that quiet forest. I know it's invisible to me, but still I strain to perceive the silent drama playing out amongst the roots beneath my feet. I gaze down at the knotted wood and try to imagine, in my mind's eye, the tendrils of connective fungus binding this forest together, carrying messages and nutrients to and fro. I can't stay for long, though. I have too much ground left to cover on my search. One of Darwin's most controversial arguments — that morality can evolve via natural selection over the millennia — lies just ahead.

9. The Caws of Culture

> "Englishmen rarely cry, except under the pressure of the acutest grief; whereas in some parts of the Continent the men shed tears much more readily and freely."
>
> — Charles Darwin, *The Expression of Emotions in Man and Animals.*

> "Besides love and sympathy, animals exhibit other qualities connected with the social instincts which in us would be called moral."
>
> — Charles Darwin, *The Descent of Man.*

1836. He is finally home. Charles Darwin steps off the Beagle a changed man. Crate upon crate of specimens and fossils are offloaded from the ship's hull– they're headed straight for the museum. While battling disease and continual seasickness on the journey back, Darwin has compiled massive volumes of research notes. Some of these notes were already sent to his mentor Henslow back in Britain, who has helped publish them in his absence. Darwin's reputation is now on the rise, and this is more than enough for him to ditch the idea of the priesthood and instead become a full-time naturalist. His father agrees, providing Charles with some seed funding to become a London Geological Society Fellow. Back from his five-year odyssey and never to leave Britain again, it's time for the real work to begin.

The next two years are a frenzy of research and writing. Darwin collaborates with some of the brightest minds in

England, and they find something incredible in the Beagle's collection: the array of birds brought back from the Galapagos, which Darwin thought made up quite a few different bird families, are all ground finches. Indeed, each of the small Galapagos Islands harbors a type of finch with a distinct beak. An idea already developing in Darwin's mind is beginning to take shape: species can undergo "transmutation,"[15] diverging over long stretches of time. These birds, along with the thousands of dead animals, fossils, skins, and pages of research notes that Darwin has brought home — will serve as the foundation for his theory of evolution over many more years.

His health is beginning to deteriorate. Heart palpitations and stomach issues drive him into the Scottish highlands, working and writing away from the noise and smog of London.[16] At the same time, a man of Darwin's station is expected to marry. He composes a highly scientific list of the pros and cons to marrying, and decides that the benefits will outweigh the costs. He begins to look for a "wife, that most interesting specimen in the whole series of vertebrate animals." In July of 1838, he starts courting his cousin Emma Wedgwood, a brilliant, well-educated woman in her own right. For the British upper class at the time, marrying one's cousin isn't unusual at all. By November, he proposes.

The two spend many evenings in deep, intellectual conversations, and are very much in love. There is a side to Charles, however, that he is hesitant to share. Some of his experiences during his travels– the treatment of slaves, the killing

---

[15] "Transmutation" was Darwin's prototype word for "evolution", but the two are one in the same.

[16] Some researchers today believe these issues stem from lactose intolerance, which worsened as Darwin aged.

of Native Americans, the existence of parasitoid wasps– continue to haunt his sense of spirituality. He's grown up as a Christian and has always believed in the doctrine, but his adventures have changed him. He's beginning to doubt whether Christianity– at least the way most people interpret it with an active, interventionist God– is completely accurate.

Emma is a devoted Christian. She is well-versed in theology and is a committed churchgoer. Charles doesn't want to upset her with his beliefs– in the conservative and religious culture of Victorian England, spiritual doubts are especially taboo. His father agrees, advising the young naturalist to keep it to himself.

But Charles can't bear to keep the truth from her. Emma responds, not with hostility, but largely with kindness and understanding. She admires her new husband's honesty, and tells her friends that Charles "was not like the rest of the Darwins, who will not say how they really are." Religion and God's role in the universe become recurring topics over the course of their marriage.

"May not the habit in scientific pursuits of believing nothing till it is proved, influence your mind too much in other things which cannot be proved in the same way, and which if true are likely to be above our comprehension?" Emma chides him.[17]

Charles knows she is right. At least, he wants her to be right. But he can't stop thinking about all the people who kill and enslave in the name of Christianity, and the moral implications of his own developing theory run deep in his mind.

By 1839, Charles has a son– William. He also has a book. His modified diary, the concisely-titled "Journal of Researches

---

[17] This syntax may be a bit hard to follow. Emma is essentially writing "God's existence cannot be proven or understood using science, so perhaps it's best to avoid letting science influence your faith."

into the Geology and Natural History of the Various Countries Visited by H.M.S. Beagle" earns the thirty-year-old Darwin fame and recognition.

But he doesn't enjoy the attention. In 1842, he retreats further and further from the limelight, moving permanently to the country with Emma and his young children.

All the while, the concept of natural selection is tantalizing. He has a full conceptual draft of the theory in his personal notes, but is well aware of what publishing it could do to his family and reputation. He keeps his radical and developing theory hidden from the public eye, only corresponding about evolution with close colleagues. The culture at the time is too religious, too conservative– people are being thrown in prison for atheism. Darwin does not want to join them. Publishing on evolution now would be "like confessing a murder."

Darwin is no atheist. He believes that God put in place the laws governing evolution, but he also believes that his principle of natural selection will be interpreted as heresy. A benign, interventionist God does not fully explain the origin of species. Darwin's anxiety spikes and his stomach churns while considering some of his predecessors who had radically challenged a religiously imposed status quo. Galileo had been placed under house arrest and just narrowly escaped execution for asserting that the Earth revolved around the Sun — and for not-so-subtly implying that the Pope was "simple" for believing the opposite. Another radical Italian astronomer, Giordarno Bruno, had been burned at the stake in 1600 for promoting the Copernican sun-centric model. Copernicus himself waited to publish his own life's work on the heliocentric system until after his death — a smart move.

A postmortem publication! That could work. In 1844, Darwin instructs Emma to publish the draft of his evolution manuscript should he die before it is finished. For now, the future of biology is shelved. Evolution lies in limbo, safely tucked away in the recesses of some dusty wooden filing cabinet. There it sits during Britain's politically-charged 1840s: stagnant, full of potential, waiting for the winds of culture to change.

///

When looking for the origins of morality and culture in humans, people like to begin by turning to our closest evolutionary relatives: chimpanzees and bonobos. After all, chimps share many physical and behavioral characteristics with humans, along with 99% of our DNA. Our relatedness can be a problem, though; when we look at the behavioral traits that are shared between close evolutionary relatives, it can be hard to say whether those behaviors are actually indicators for advanced sociality or just left over from a shared ancestry.

For example, Chimpanzees have facial expressions and are smart; humans have them as well and are *very* smart, so facial expressions might be needed for higher sociality. But chimps and humans are both primates, all of whom use facial expressions to communicate — and not all primates are as smart as humans or chimps. Meanwhile, dolphins do not use facial expressions at all, yet they have incredibly complex social systems.

Sticking too close to our primate cousins can lead us to miss some important details about social animals. Animals are behaviorally diverse, but we still see some patterns of behavior across species that are hundreds of millions of years of evolutionary divergence apart. Enter my favorite vertebrates: birds.

Birds are the descendants of dinosaurs. The Costco $4.99 rotisserie chicken lies in the same taxonomic clade as the tyrannical *T. rex*. Since giant dinosaurs died out long before mammoths or humans got their day in the sun, most assume that dinosaurs evolved before mammals. It's the opposite; warm-blooded mammals evolved from our reptilian ancestors long before dinosaurs. Dinosaurs and modern-day birds run on some of the most unique hardware on the planet, and they evolved their most defining features, like warm-bloodedness and powerful brains, independently. Yet, advanced cognition is eerily similar in birds and mammals —both have converged on a similar working strategy.

Roughly 300 million years ago, all of the would-be reptiles, birds, and mammals were part of a single group called the Amniotes, named for the amniotic fluid which helps provide nutrition and water to their eggs. Shortly after, the Amniotes split into the Synapsids (which later became the mammals) and the Sauropsids (which still includes lizards and birds/dinosaurs). The latter group only split off to become proto-dinosaurs in the 120 million years that followed. Brain-wise, these species were likely far from what today we rank as the smartest animals.

Despite their evolutionary distance from humans, birds like parrots or corvids are fiercely intelligent. In cognitive testing, they regularly score about the same as chimpanzees, orangutans, and dolphins, with the intellectual abilities of a five- or six-year-old human. Considering how intelligent a six-year-old can be, that's pretty astounding. Many studies demonstrate the intelligence of birds. For example, Alex the

African Grey Parrot was part of a 30-year communication and learning experiment by researcher Irene Pepperberg that ended in 2007. By 1999, Alex knew over 150 English words and could use them to distinguish between different colors and shapes. If you asked him, Alex could tell you the color, shape, and makeup of any object you put in front of him within reason. He could also distinguish between objects that were larger or smaller than one another, as well as "same" and "different" objects.

It wasn't a circus act. Alex must have understood the meanings behind his words because *he invented new ones*. In my favorite instance, Alex was shown an apple, for which he had not learned the word. When asked what it was, Alex called it a "banerry," combining the words he knew for "cherry" and "banana." Germanic languages like Swedish, Dutch, and German do something similar: when they need a new word, they often just smash two or more existing words together. This is why the longest word in the German dictionary, meaning "motor vehicle indemnity insurance" is *Kraftfahrzeughaftpflichtversicherung*— instead of, you know, something reasonable.

Alex is also the first and only non-human animal on record to have asked a question. Apparently, while looking at himself in the mirror, Alex asked, "What color?" The researchers responded, "Grey." It might not seem like much, but the fact that animals like Alex can recognize themselves in the mirror and even ask *questions* about themselves speaks to the existence of a mind, capable of understanding the world in a fantastically insightful way.

Crows and ravens are likewise gifted. They can mimic human speech, just like parrots, and even solve complex puzzles requiring a grasp of physics. One water-displacing experiment for crows is set up by placing a crow into a cage containing a water-filled test tube. A piece of meat floats just at the top of the tube. Some pebbles are scattered about the cage as well. The crow wants the meat, but the meat is too far down in the tube for it to reach. After a bit of thinking, the crow grabs a pebble and drops it into the test tube. This raises the water level just a bit. The crow then continues its avian Archimedes act until the water level is high enough for it to reach in and grab the floating meat with its beak. Eureka!

Above: these clever corvids are master puzzle-solvers.

I could — and often do — go on and on about how brilliant birds can be, how some create and modify their own tools, how they build dazzling and aesthetically pleasing structures, how they dance to human music,[18] invent their *own* sexy dance routines, and so on. It turns out that corvids don't just compare with humans in our ability to use tools and grasp the laws of physics. They may also exhibit something we long thought unique to ourselves: culture.

---

[18] Try Googling "Snowball the Dancing Cockatoo."

In the early 2000s, at the University of Washington, a scientist named John Marzluff and his team conducted an experiment involving crows and masks, testing whether crows could tell one human face from another. They used two different types of masks for the experiment — one mask was of a caveman's face, and the other was then-Vice President Dick Cheney. When researchers wore the Dick Cheney masks, they left the crows alone. When they wore the caveman masks, though, they captured some of the crows and took their measurements.

Crows loathe being handled against their will. Their captors were doubtlessly met with bouts of indignant caws and attempted bites. It wasn't long before the researchers saw a corresponding change in the crows' behavior. Whenever a researcher walked by wearing the Dick Cheney mask, the crows didn't mind, even if the liberal college students did. Whenever a researcher wore the caveman mask, the crows went ballistic. Researchers in caveman masks were often scolded and occasionally dive-bombed, a classic anti-predator behavior in many bird species. There was no doubt about it: the crows *recognized human faces* and changed their behavior depending on whose "face" it was.

It gets stranger. Word of the caveman researchers and their nasty habit of putting crows in cages got around in avian social circles. In fact, five years after the initial experiments had ceased, the caveman-mask scolding behavior had spread to crows across a 25-mile radius of the University of Washington campus, where the experiments had first taken place. Crows that weren't even *alive* when the experiments happened were still scolding anyone in a caveman mask. The campus crows had started a ritual that lived on after them.

How is that possible? Crows don't talk to each other per se. They weren't able to add "caveman-looking people" to their list of unholy humans in the Crow Testament. Instead, the crows learned by observing the behavior of their fellow crows. See your family and friends screaming at people in caveman masks, and you learn to associate those masks with danger. If you're a crow, *even if you haven't experienced the danger yourself,* you can still be conditioned to respond to that danger.

Above: two crows keep a lookout from atop a building. Photo by Ryan Hodnett. Accessed via Wikimedia Commons.

This is socially acquired fear. From an evolutionary standpoint, it's an ingenious adaptation. In order to establish a threat response, many animals either have to go head-to-head with danger or else evolve an inherent naturally selected repulsion to certain organisms or natural formations. Some animals do this to a lesser extent with longtime predators, but these Seattle crows learned to avoid cavemen just from being captured and released over a matter of months. Often, if you are exposed to a real threat, you don't get a chance to learn from the encounter. You're finished. Knowing what's dangerous is a matter of life and death, and thus best learned from others.

The crows from Marzluff's experiment showed a learned behavior that was repeated, spread, and passed down to a succeeding generation. We can see the resemblance to humans. The Seattle crows were exhibiting something like culture, maybe even a tradition.[19]

In the opening of *Fiddler on the Roof,* Tevya muses, "You may ask 'How did this tradition get started?' I'll tell you. ... I don't know. But it's a tradition!" Likewise, younger crows don't exactly know why they caw at the humans in caveman masks; they've never been attacked by one, but they do it anyway, because that is what their parents did and what their friends do. It's peer pressure. An animal doesn't need to grasp *why* a learned behavior is useful for it to be beneficial. That's the beauty of cultural learning: animals learn useful behaviors

[19] By the way, not all traditions and cultural ideas are beneficial to survival and reproduction. In 2014, researchers found that a group of chimpanzees had started a tradition of wearing a piece of grass in their ears as a sort of fashion statement. Can we trace that directly to reproductive success? Maybe not; maybe it was just an aesthetic choice.

without even needing to experience the dangerous contexts that gave rise to them.

Most of the world's smartest animals share some traits. They are usually highly social, exhibit some form of tool use to the extent that their appendages allow, and employ complex systems of communication.[20] They are also often generalists, hunting and foraging for a variety of different foods, rather than specialists. Even though high-powered bird brains evolved separately from high-powered mammal brains, corvids and parrots exhibit a tremendous amount of behavioral overlap with primates and whales, the two smartest families of the mammal world. These are the brains of long-lived generalists: powerful, socially-keen, and flexible. A more high-powered brain allows animals to adapt their tactics at a rapid pace, much faster than the speed of evolution. Clever organisms use culture as a way to learn and teach about the world around them, and to change in response.

Like humans, crows may even hold something like a funeral or death ceremony. The science behind this isn't fully understood yet, but researchers have watched crows gathering around dead members of their species. As more and more crows arrive, they form a circle around the deceased, then begin to caw loudly to onc another. Finally, the birds go quiet and stand vigil, until finally flying off one by one.

Crows could hold this "ceremony" for practical purposes. Maybe they just want to learn about threats in the area, but I wonder: do crows feel sorrow when they see another crow dead? Is there some aspect of the gathering and cawing that

---

[20] That is, these animals all do better in problem- and puzzle-solving tests, which could admittedly lead to some biases because these were also designed by social animals: humans.

might be cathartic, especially if the body is of a former member of the flock? We know that crows and ravens can make lifelong friends and even recognize them after decades apart. It stands to reason that they'd feel pain at losing a family member or friend. As we are just now beginning to realize, animals have the capacity for far more emotional depth than we ever imagined.

Elephants, another cognitively gifted animal, hold even more intimate ceremonies for the dead. When elephant herds encounter a corpse, they often fall strangely silent. Members of the herd begin to look over the corpse, passing their trunks gently over the body. Even if they do not recognize the corpse as a dead relative, they often stay with the body for some days.

When elephants do recognize family members, the grieving is much more intense. In his book *Elephant Destiny*, journalist and historian Martin Meredith recounts the following scene:

> The entire family of a dead matriarch, including her young calf, were all gently touching her body with their trunks, trying to lift her. The elephant herd were all rumbling loudly. The calf was observed to be weeping and made sounds that sounded like a scream, but then the entire herd fell incredibly silent. They then began to throw leaves and dirt over the body and broke off tree branches to cover her. They spent the next two days quietly standing over her body. They sometimes had to leave to get water or food, but they would always return.

Elephants also recognize death when it takes place in front of them. The researcher and conservationist Cynthia Moss corroborates this in her own book, *Elephant Memories*, recounting the moments after an elephant was shot by a poacher. She writes:

> Her family tried everything to rouse her, kicking and tusking her, and Tallulah even went off and collected a trunkful of grass and tried to stuff it into her mouth ... They gave up then but did not leave ... By nightfall they had nearly buried her with branches and earth ... Then they stood vigil over her for most of the night and only as dawn was approaching did they reluctantly begin to walk away, heading back toward the safety of the park.

Again, a practical explanation is that the elephants were examining the bodies of the dead to learn about threats. Perhaps they throw leaves and dirt over the body to mask its scent and avoid attracting predators. Maybe they stand over the body to mourn the loss of a loved one, remembering the deceased, until they find some measure of closure and move on. There's no reason why ceremonies involving the dead can't serve all of these purposes. That's the power of culture.

I could probably discuss hundreds of examples of cultural learning in non-human animals. Chimpanzees learn to fish for termites with sticks, and they teach the technique to their friends and children. Some whale pods are led by elderly matriarchs who remember the migration routes to long-used feeding grounds (not unlike elephants). Songbirds sing in regional accents that they pick up from their parents and neighbors. And so on and so forth. Culture affects the way we interact with the outside world and within our social groups, and those interactions dramatically affect our chances of survival and reproduction.

From the day we are born, humans are fully immersed in culture. We also make the distinction between good and bad, or right and wrong, behavior, and we call this process of distinguishing "morality." As with many of the social species depicted in this book, morality is transmitted via culture, allowing a species to exist in more harmonious and more effective groups over time. Even so, how we got to this "higher" moral thinking — a question much older than Darwin's attempts to explain it — is still far from fully understood. What makes our ways of seeing the world, and each other, "moral?" Are we different from a dog that breaks loose to chase a squirrel, or a cleaner wrasse punishing a greedy partner? The complexity of culture in larger-brained social animals should clue us in — the answer lies in the mind. To understand where moral behavior comes from, we first have to understand emotion, empathy, and the way human brains are built, first and foremost, to navigate the complexities of social life.

10. Climbing the Cognitive Ladder

> " … The large size of the brain in man, relatively to his body, in comparison with that of the gorilla or orang, is closely connected with his higher mental powers … "
> — Charles Darwin, *The Descent of Man*

Brains are complicated. Brains are so complicated, in fact, that calling them "complicated" is like calling the inside of the Sun "warm." In cranially gifted animals like humans, much of that brainpower goes toward assessing social relationships. In the context of behavioral evolution, we get to see why.

Primatologists tell us that chimpanzees and other Great Apes play complex political games with one another, constantly mapping out the social dynamics of their groups and assessing their own ranks within them. Even more "peaceful" great apes, like bonobos, have rich matriarchal societies in which a parents' rankings and friendships can have a big impact on their children's own outlook. To play the Great Ape social game well, you need a powerful mind that can keep track of faces, relationships, past behaviors, friendships, rivalries, and more. All of this information goes toward building the right alliances and protecting oneself from the wrong enemies. It's middle school on steroids.

Like we've seen in previous chapters, large social groups are born from mutualisms, where everyone benefits from a partnership. Group members have a reason to stick together. Groups work most *effectively* when everyone cooperates. The problem is, of course, that even though groups that cooperate do better than those that do not, individuals that cheat the system can sometimes do even better than the honest members of their group. Since natural selection works at the *individual* level, we inevitably see "cheaters" emerge and take advantage of the generosity of others.

Above: this Bonobo is likely part of a large social group, with plenty of relationships to keep track of. Photo by Pierre Fidenci. Accessed via Wikimedia Commons.

Cheaters sometimes win. Too many cheaters in a social system, though, and the entire structure becomes pointless. Social organisms need a system in which they can recognize offenders and punish them. It's a delicate balancing act — each organism wants to maximize the number of offspring it has and game the system as best it can, simultaneously punishing others who do the same. What's the best strategy to adopt? It turns out that there's a mathematical answer to the problem of cooperation.

Robert Sapolsky, a behavioral biologist, shows the economics of cooperation by starting with the "Prisoner's Dilemma." The premise is something like this: you try to break out of prison with your fellow inmate Jeffrey, and the two of you are captured during the attempt. You are then interrogated separately, in soundproof rooms, and both Jeffrey and you are given the same choice: either rat out your friend, or keep quiet. If neither of you rats the other out, you both serve a light sentence for your crime. If both of you expose the other, you both serve a pretty long sentence. If you rat Jeffrey out but he doesn't rat you out, you serve almost no jail time. If Jeffrey rats you out but you don't rat him out, you serve the maximum sentence. The question is: to rat or not to rat? Take a second to think about what you'd do in this scenario.

If you said "reform the prison system," I bet you're fun at parties. Assuming this behavioral dilemma is happening just *once*, the best choice is actually to rat Jeff out. You avoid the worst outcome, and there is a chance of you getting the best one if Jeff is nicer than you. It sounds counterintuitive, but in computer simulations, always "cheating" is the best strategy if the game lasts *only one round*.

Real social life involves more than one round of cooperating or cheating. It's many, many rounds. Play the game over and over with multiple metaphorical prison breaks, and cheating is no longer the most effective way to live. People who *never* cooperate will never get any benefits from cooperation, because nobody wants to work *with* them. Likewise, if a person *always* cooperates, they're probably bound to get screwed over by a cheater. So what changes in a multi-round game?

The introduction of more than one round to the Prisoner's Dilemma opens up a strategy known as "tit-for-tat." Basically this means that if Jeff cooperates with you this round (tit), you cooperate with him the next round (tat). You could also call this "reciprocal altruism." If Jeff cheats you this round, you cheat him in the next. If, after that, he goes back to cooperating, you go back to cooperating, and vice versa. The idea of tit-for-tat is this: "I scratch your back, you scratch mine, but if you stab me in the back, you best believe you're next if I get the chance." A kindness for a kindness. An eye for an eye.

There can be more layers of complexity, like "forgiving tit for tat," which forgives a transgression after a while to see if switching to cooperation will change the other's cheating behavior. Sometimes it does. It turns out forgiveness can be useful, too. If you show "forgiveness" and the other person keeps cheating, though, you shouldn't keep forgiving him for long. In commandment form: "First, do unto others as you would have them do unto you; then do unto others as they do unto you." Tit-for-tat rewards cooperation, punishes cheaters, and is forgiving. In short, it's a very dynamic behavioral strategy and it's easy to see why so many animal species, including humans, use this kind of behavior.

Vampire bats, of all species, are some of the most well known reciprocal altruists in the animal kingdom. Since they survive only on blood, their feeding habits are very hit-or-miss: either you get a drink that day or you don't. When vampire bats go hungry, their neighbors will share the blood they found that day by regurgitating it. That kindness isn't free. Blood donor bats have a long memory. They expect that the next time the situation is reversed, the bats they help will help them in return. The vast majority of the time, that's what happens; vampire bats can trade off food-sharing with trusted partners for years. If bats get stiffed, though, there's a good chance they'll withhold their charity the next time around. That's tit-for-tat.

Vampire bats employ even more layers of sharing complexity, like sharing *more* food with someone they weren't able to help before, or being more willing to share with someone with a good track record of reciprocating. Vampires pick their friends carefully and keep track of their favors and debts — all that social complexity might be one reason why vampire bats have a much larger neocortex than their less social fruit bat relatives.

As Dr. Christopher Boehm notes, humans are much like vampire bats, whether those humans live in sedentary agricultural communities or in hunter-gatherer communities.

> "If bad luck arrives, band members will, within limits, help others in the band according to need- but even though any person of good or acceptable moral standing is automatically eligible for these indirect reciprocity benefits, a few very large beans will, in fact, be counted. In general individuals whose track records are unusually generous may receive more help than those who have been stingy. And as for people who have a long record of being outlandishly and immorally selfish or lazy ... When personal trouble strikes, and with only their kin likely to support them, such individuals may wish they followed the golden rule."

Darwin believed as much when it came to human behavior. He observed reciprocity as a "universal" constant in human cultures, which served as a founding principle of society. Robert Richards writes that

> Darwin noticed, of course, that people ... formed contracts to respect the person and property of others, provided they received the same consideration themselves; they acted, in our terms, as reciprocal altruists. Darwin also observed that his fellow creatures glowed or smarted under the judgments of their peers; accordingly they might at times practice virtue in response to public praise rather than to the inner voice of austere duty. Yet individuals did harken to that voice, which they understood to be authoritative, if not always coercive."

Economic experiments involving tit-for-tat behavior were run on computers or done with small groups of humans or other animals, and they help to show us how what we'd call "morality" arises from behavioral evolution. To reiterate, social groups rely heavily on honesty. It therefore makes sense to see honest cooperation as "right" and cheating as "wrong." There's a reason it angers us so much when someone cheats in an athletic event, or at a video game, or on their taxes.

Humans, along with other social animals, are hardwired to spot cheating in their social groups and punish the cheaters, sometimes severely. In real life, "cheating" includes any behavior that breaks a system of trust, like theft, adultery, picking Rainbow Road in Mario Kart, or price-fixing. Animals that punish cheaters and reward cooperation will pass on more copies of their genes in the long run, because they'll survive and reproduce better in amicable, harmonious[21] groups than in

---

[21] "Harmonious" might be too strong a word; maybe it's more like "not actively at one another's throats."

groups constantly wracked with internal turmoil. Teamwork not only makes the dream work; it also makes your genes work.

Take another classic moral commandment: "Thou shalt not kill." Our social conditioning is not the only reason that we avoid murder. Even without a mandate from the heavens, there's a pretty good biological reason not to go around killing other people in your group.

Let's start with the most obvious: if you live in a group with someone, there's a decent chance that person is closely related to you. In other words, the person probably shares your genes or shares genes with a niece or nephew or half-sibling. We already talked about kin selection, which shows us that even if you don't reproduce, your gene-carrying family members still can, and it behooves you to keep them safe for that reason. If you go around eighty-sixing the members of your tribe, you're not doing anything to benefit your genes.

Some societies aim to bypass divisions of kinship altogether, by trying to make groups feel like a family. Soldiers are sometimes called "bands of brothers," and that's an intentional move. Boehm also writes about societies "tweaking" human nature: "Among unrelated families in a band, such 'tweaking' is needed because human altruism is weak compared to nepotism." By creating a strong feeling of kinship between people, even if they're not actually related, you can increase the likelihood that they will work together, cooperate, and make sacrifices for one another. I'd be willing to help a stranger, but even more willing if I thought of that stranger as my brother.

Next on the long list of reasons why you shouldn't kill the people you live with: we need one another. The world is filled with parasites, predators, Mean Girls and competitors. If you hunt and forage and defend your territory as a team, you need all hands on deck. Killing one of your own tribe puts you at an immediate disadvantage, even if that person isn't closely related to you. In the animal world, even intense rivalries between group members rarely result in death.

For example, many animals are "tournament species," which means that members of one sex (usually males) compete against one another for access to mates. In these species, a small handful of males account for 95% of the matings, while those on the bottom rungs do not get laid at all. The difference in success can be so stark that it's evolutionarily better for most male turkeys to help their stronger, larger brothers mate than to try and find a mate on their own.

Crucially, fights between contest maters, though brutal, almost *never* result in death. Deer and goats will butt heads with their antlers and horns, kangaroos and horses box one another for dominance, and chimpanzees get involved in intense, screaming political matches to gain social dominance, but none of these species engage in rituals that regularly kill members of their own group while competing for mating access. It's not worth it.

If you're a fan of Biblical loopholes, you may point out that if translated from the original Hebrew, the sixth commandment *technically* reads "Thou shalt not *murder*." You, my loophole-finding friend, would be right, and that's no small difference.

Even in non-human animals, it can be totally acceptable to kill a member of your own species, so long as that member isn't kin. Ants of the same species go to war all the time, slaughtering entire colonies, but in-fighting within a colony, barring some political upheaval like two queens splitting loyalties, is extremely rare. Chimpanzees very seldom kill members of their own group, but they will raid rival groups. Some of these raids result in systematic killings. The "Gombe

Chimpanzee War," a conflict that took place between two chimpanzee groups from 1974 to 1978, left 11 chimps dead. As world-famous primatologist Jane Goodall said, "if [chimpanzees] had had firearms and had been taught to use them, I suspect they would have used them to kill."

Remember, individuals don't act for the good of the species; *they act for the good of their genes.* Some organisms, like Genghis Khan, try to advance their genes by plundering resources and killing off neighbors. Today, 1 in 200 men may be descendants of the colonizer. Sometimes this strategy works, other times it does not. In Samuel 15:18, God says, "Go and completely destroy those wicked people, the Amalekites; make war on them until you have wiped them out." The Bible's distinction between murder (wrong) and killing in a time of war (heroic) is congruent with our biological background.

Biology reinforces our moral revulsion to violence against women. As primatologist Frans de Waal points out in his book *Mama's Last Hug,* male apes often fight each other for mates while following "standards of engagement" — even if they're capable of it, they tend to avoid lethal force. For most tournament species, harming another male makes sense during competition, but harming a female doesn't. De Waal writes:

> ... the fighting rules have hardly changed. They evolved over millions of years and are far too ingrained to be thrown out. A male generally curbs his physical power while confronting a female. This is as true for horses and lions as it is for apes and humans. These inhibitions reside so deeply in our psychology that we react strongly to violations. In the movies, for example,

it's not terribly upsetting to see a woman slap a man's face, but we cringe at the reverse.

Evolved behavior whispers to our moral compasses. Human groups tend to condemn male-on-female violence far more harshly than they do male-on-male violence or female-on-female violence.

I saw a fair number of fights as a high schooler. I got very close to being in one after some baseball players from another school recognized me as the pitcher from a previous, heated game. Neither of us helped defuse the situation.

It's amazing just how "ritualized" a fight can feel. There's always an unspoken protocol — the crowd forms a ring around the fighters, almost always two males, who begin by posturing and hurling insults. Once physical contact starts, fists fly. The winner is the first person to knock the other down or out, and stepping out of the circle means you're done. Unfair fights more quickly end with both combatants getting dragged apart. If there were ever a time when these kinds of fights were evolutionarily beneficial, I think it's long since passed. Today, sports are a much more acceptable form of ritualized physical competition. Unsurprisingly, a large subset of humans find high-performing athletes sexually appealing.

Many human societies have male or female-dominated structures, similar to those of chimpanzees and bonobos. It's easy to understand why humans allow ritualized violence under the right circumstances. Even with all of this evidence from evolutionary biology, there's a nuance to some of our moral beliefs that doesn't fit the bill. Many humans, despite having evolved within systems of ritualized violence, condemn violence of any kind and swear to be pacifists. Others make vows not to have any children,[22] instead serving a community or a god. How do we explain these behaviors? Well, we're getting there. It turns out that we're missing a key ingredient of sociality in cognitively advanced animals. We are missing empathy, and the all-important Theory of Mind.

[22] Such as celibate priests, nuns, and members of the anti-natalism movement.

11. A Mind for Others

"It must not be forgotten that a high standard of morality gives but a slightly or no advantage to each individual man and his children over the other men of the same tribe, yet that an increase in the number of well-endowed men and an advancement in the standard of morality will certainly give an immense advantage to one tribe over another."

— Charles Darwin, *The Descent of Man*

"... for the shield may be as important for victory, as the sword or spear."

— Charles Darwin, *The Origin of Species*

Three and a half years of hell. Three and a half years of crippling stomach pain, long nights up reading manuscripts, changing paragraphs, and revising, revising, revising. But it is done. The Origin of Species is finished.

Darwin could not have completed the book without the support of his friends and colleagues, and was perhaps further bolstered by the fact that another researcher– Alfred Russel Wallace, had recently developed a theory about organisms changing over time that looked very, very similar to his own. Both researchers had their work presented at the Linnean Society in 1858– but at the end of 1859, the catchily-titled On the Origin of Species by Means of Natural Selection, or the Preservation of Favoured Races in the Struggle for Life is finally released to the ravenous public.

On one hand, it's like a great weight has been lifted off Charles' shoulders. On the other, he is getting the hell out of dodge and living somewhere quiet in preparation for an inevitable firestorm of controversy. He does not have the stomach for what is to come.

And of course, the firestorm comes. Reviews of the book, many of which are from Charles' mentors and colleagues, range from collegial to scathing. Natural selection is lampooned by many of the world's most famous thinkers, venerated by others. By far, the most ridiculed sentiment is that humans descend from primates– a statement that Darwin very deliberately left out of the Origin of Species but was interpreted as such anyway.

Unsurprisingly, the strongest critiques come from the devout. Criticism of the book on religious grounds begins to slowly trickle in, and then it becomes a flood. Natural Selection is called "a brutal philosophy — to wit, there is no God, and the ape is our Adam." Commentators claim that evolution is "dangerous science" and "an attempt to dethrone God." One critic goes so far as to say, "If the Darwinian theory is true, Genesis is a lie, the whole framework of the book of life falls to pieces, and the revelation of God to man, as we Christians know it, is a delusion and a snare."

Samuel Wilberforce, the bishop of Oxford at the time of The Origin's publication, declares that Darwin has "a tendency to limit God's glory in creation" and feels that the idea of natural selection was "inconsistent with the fullness of His glory ... a dishonoring view of Nature." Wilberforce was a renowned public speaker, and so prominent a figure's coming out against Darwin's theory did nothing to stem anti-evolution sentiment.

In 1860, the famous Oxford Huxley-Wilberforce debate takes place, in which "Darwin's bulldog" and ardent theoretical defender Thomas Henry Huxley faces off against bishop Samuel Wilberforce. The two men are rhetorical powerhouses– accomplished speakers, and the topic at hand is "descent", another of Darwin's terms for evolution. The men engage in a furious back-and-forth, both winning points from a rapt audience and speaking with passion. It is blazing hot indoors. Many men rise to support one side or the other. FitzRoy, Darwin's captain, holds aloft a Bible and proclaims "had I known what I know now, I would never have taken [Darwin] aboard the Beagle."

At one point Wilberforce apparently asks Huxley whether his ape descendants were on his grandmother or grandfather's side. People in attendance claim Huxley replied "If I had to choose, I would prefer to be a descendant of a humble monkey, rather than of a man who employs his knowledge and eloquence in misrepresenting those who are wearing out their lives in the search for the truth." The crowd goes wild. A woman faints. Every speaker claims victory by the end, and they all leave the lecture hall in good spirits to go drink and dine.

And in the midst of it all, Charles Darwin's voice is conspicuously absent. Few have heard from the man in months, but this is not unusual. The father of evolution is shy and reclusive. One wonders how the debate might have gone differently if he were there. If the people in that room knew what Charles had gone through, if they understood what he had experienced, perhaps they would see his reasoning in a different light. Perhaps they would understand why, to Darwin, natural selection is essential for making peace with the machinations of God.

///

Among social animals, knowing what another member of your species is thinking and feeling can save your life. A major rung up on the ladder of social cognition is known as the "Theory of Mind," or the capacity for an organism to recognize the mental state of another. Theory of Mind is vital for humans. As our brains develop, so too does our capacity to recognize the humanity — or inhumanity — of others.

The first step to developing an individual's Theory of Mind is recognizing the "self" as distinct from others. Many animals recognize themselves in the mirror. We know this thanks to a researcher named Gordon Gallup, who ran a series of experiments with monkeys, apes, and mirrors. Gallup put small red dots on the faces of sedated primates, and then added a mirror in their cages. This didn't do much to the monkeys, but when the chimpanzees, bonobos, and gorillas awoke, they almost always looked at the mirror and reached up to touch the red dot on their faces. The apes would only do so if they understood that they were seeing *themselves* in the reflection. Otherwise, they'd interact with the reflection in the mirror, not their own body.

It takes some time for human babies to reach that point. Research in pediatric psychology suggests that humans only form "ego boundaries," the separation of the self from others, past a certain point of development as a toddler. Once you recognize that nobody shares your brain, you can recognize that others may have information you don't. A related behavior is called "gaze-following," whereby individuals will look in the direction that someone else is pointing or looking toward. Little kids revel in pointing in some direction and saying, "Hey, what's that over there? Made you look!" It works for a reason.

Gaze-following exists in primates, dogs, ravens, and even goats. A gaze might signal to one of these animals: "Hey, that goat's getting information about the world by focusing in that direction. I'd better check out what's so interesting," or something along those lines. Next comes the true test of Theory of Mind: being able to recognize that everyone else's mind is *unique.*

Imagine you're playing a game of poker. The point of poker is to win all of the available money by building hands of cards, thereby making everyone else at the table upset. You can bet money on your hand, call the bet of another player, raise, or fold. There's a lot of "mind-reading" involved. You need to assess what the other players might have in order to understand whether betting is a good idea. You can try to bluff, but that takes us higher and higher up the ladder of Theory of Mind. Here are some of the "rungs" along the way.

1. **I know that I know.** Here's that separation that older toddlers make between their own experiences and the experiences of others. I recognize that only I can see the cards I have in my hand. That is, unless someone is cheating. Otherwise, nobody else knows for certain what my cards are.

2. **I know that you know.** You can think and process the same way I can. Your momentary grimace when you look at your cards tells me something: you don't like your hand. I can see it in your eyes, which are wandering to the two of clubs on the table. Hmm ... do you have a pair of twos? There's also a queen on the table. Maybe

you'll bet, to make people think you have a pair of queens instead.

3. I know that you know that I know. Hang on. I'm not the only one with Theory of Mind. You have it, too. This is what researchers call "Secondary Theory of Mind." What if your grimace when you looked down at your cards was *intentional*? What if you wanted to give off a verbal signal that you don't like your cards to make me overconfident? That way, I will think that I've figured you out, when in reality, I've just been doing what you wanted me to. You sly dog. You've been playing me like a fiddle this whole time. Well, I know that you know that I know that you know that I know ...

... and on and on and on. When two beings are capable of "mind reading" and know that the other is, too, mind-reading games like these can get heated. As we'll see, this level of complexity takes quite a bit of brain development, even in humans.

Here's a popular Theory of Mind test that's done in front of children. It has its problems, but works as a simplified example. A researcher puppets two dolls– "Sally" and "Anne." The dolls play with a ball. After a while, Sally puts the ball in a bag and "goes away" (the researcher takes Sally out of view). While Sally is gone, Anne goes into the bag, takes the marble, and moves it to a box. Sally returns. Where will Sally look for the marble?

If the child being tested is under four years old or so, they will usually point to the box where the marble is, not the bag. The kid knows the marble is in the box following a question from the researcher, but doesn't seem to understand that *Sally* wouldn't know this herself. If you ask a child over the age of four the same question, they most often point to the *bag*, not the box. They understand Sally would *not* know her marble was moved. In other words, people four and up are able to put themselves in the minds of others.

Attributing this "false belief" to someone else is a huge step up in terms of social smarts. With this knowledge, the world immediately becomes a more complex place: people don't have the same information about the world that you do, and because of this, they can react differently to the same events. A three-year-old watching the end of *Romeo and Juliet* (and having been able to grasp the rest of the play) would be wondering why Romeo is so broken up over Juliet's feigned death. She's not dead, after all, just asleep from the potion that the good apothecary gave her. A neurotypical four-year-old, on the other hand, could understand that, even though Juliet isn't dead, Romeo *thinks* that she is. Shakespeare's iconic play is a true tragedy of false beliefs.

For a while, humans believed that they were alone in using a high level of "mind-reading." Mentally putting yourself in someone else's shoes, we believed, would be too hard for other animals — they didn't even wear their *own* metaphorical shoes. Or so we thought.

Scientists collaborating from Duke University and Kyoto University set up something of an ape "soap opera" for their chimpanzee, bonobo, and orangutan subjects to watch while sipping on orange juice. A long straw for the orange juice is used to keep their heads in one place. As the subjects watch from a screen, their eye movements are tracked to see where their attention is directed. It's important to track eye movement because you can't really ask a great ape, "Where would Sally look for her ball?"

The researchers showed the primates a couple of scenes. In one, a human dressed in a gorilla suit dubbed "King Kong" steals a rock from a man, hiding it in one box or another, just like the bag and box in the Sally and Anne experiment. The man sees where King Kong has hidden the rock, but then King Kong scares him out of the experiment room. While the man is gone, King Kong switches the location of the rock from one box to the other. Kong leaves and then the man comes back into the room. Where will he look for the rock?

According to the study featuring fourteen chimps, nine bonobos- and seven orangutans, the majority of the Great Apes looked not at the box where Kong had moved the rock, but where the man believed the rock to be. In a similar set-up featuring a human with a big stick chasing after a hiding Kong, test subjects still eye-tracked toward where the human believed Kong to be. The results of this study were unheard of — here was direct evidence that other animals could exhibit a similar level of cognitive understanding as a young human. On logic and reasoning tests, Great Apes do about as well as human five-year-olds. We have even more in common with our closest relatives than we thought.

Theory of Mind doesn't just deal with what others are thinking. It also can describe what others are *feeling*. Empathy — imaginatively feeling the emotions of another being — is everywhere in the animal kingdom. Dogs comfort their owners and packmates when they display signs of sadness or distress. Chimpanzees console friends by hugging after difficult physical altercations or social struggles. Ravens cheer up friends who've been in fights, preening each other's ruffled feathers.

Even mice and rats show empathy. In one 2006 study, mice that observed a cagemate in pain developed a heightened sensitivity to pain themselves, when compared with mice that didn't see a hurting cagemate. In a 2011 study at the University of Chicago, researchers devised an experiment that gave rats a choice between freeing a trapped rat in distress and eating a piece of chocolate. Most of the time, the rats saved their alarm-calling cagemate *first*, and then went for the chocolate, which they preferred to share with the other rat. Quite the moral rodents.

The rats were clearly acting to save their neighbor from distress; they weren't particularly interested in opening the traps when there was a toy rat or the trap was empty, but when there was another living being in there, they sprung into action. Was it annoying alarm calling that drove the rats to free their companions, or what we might call "compassion?" If the latter (which the researchers favor, given the frequency of alarm calls), then our search for the biological source of morality has paid off. If even rats — no offense to rats — can exhibit deliberate, altruistic, empathetic behavior, what might some of the smartest animals be capable of?

In his book *Behave: The Biology of Humans at their Best and Worst,* Dr. Robert Sapolsky talks about how empathy and Theory of Mind can intermingle, going from "feeling someone's pain because you *are* them, to feeling *for* the other person, to feeling *as* them." Small children lack the ability to distinguish the pain of their mothers from their own.

Two rats - highly empathetic beings. Photo by Sarah Fleming. Accessed via Wikimedia Commons.

That's a different kind of empathy than seeing someone in pain and cringing yourself. In fact, young children can't even distinguish between intentional and unintentional harm. Empathy involves feeling another's pleasure or pain, but only later in development does the context of these feelings start to factor in.

It's common knowledge that dogs exhibit empathy for their humans. Dogs are so in-tune with the emotional states of their humans that there are even multiple reports and adorable internet videos of uninjured dogs limping alongside people using crutches. The dog, despite not limping before seeing this behavior from the human, begins limping as well. Perhaps the empathetic capacity of dogs is close to that of very young children. Dogs could lack the ability to separate the experiences of their owners from their own, even if they are not in any physical pain themselves. It would make sense in the context of this "mirroring" behavior; if the dogs were able to separate themselves from their humans, they would probably focus on comforting the human rather than limping alongside.

Then again, maybe dogs are just limping to copy the human's behavior. Everyone wants to fit in. If the leader of your pack is limping, maybe it just makes sense to limp along. Well, beautiful theories are regularly destroyed by ugly facts. The best way to know is to experiment.

Why does empathy matter evolutionarily? Empathy hurts. It can be physically painful to see someone else, especially someone close, in pain, and too much exposure to others' pain can result in the phenomenon known as "empathy fatigue" or "compassion fatigue," a kind of emotional exhaustion where the brain is overwhelmed by empathy-arousing stimuli. This has been known to affect therapists, trauma workers, nurses, and other front-line healthcare workers. In one 2010 study published in the *Journal of Emergency Nursing*, some 86% of emergency room nurses exhibited some degree of compassion fatigue. During the recent Coronavirus pandemic, that number shot even higher. Why do humans have such a strong sense of empathy? Why do I cringe when someone else stubs a toe, or when a guy on TV pours milk and cereal all over himself on the New York subway? Why did I cry like a baby at the start of *Up*? Why do I care about other people's feelings?

In the earlier example of crows and culture, we saw crows learn to associate caveman masks with danger and distress. Part of what made this association so effective was crows *feeling* the emotion of the crows around them. Emotional contagion is undeniable; collective empathy can lead one person's feeling a certain way to "infect" an entire room. Laughter is contagious, and so is anger.

Some years back, I was at the movies, watching *Knives Out* with my extended family, when I witnessed a classic case of emotional contagion. Daniel Craig's character was going on some bit about a hole within a donut hole and my aunt, quick to laugh as she is, fell into hysterics. She was the only one laughing in the entire theater, and uncontrollably at that. After about fifteen seconds of laughing into the void, a few people behind her started to laugh. Then a few more. Pretty soon the entire theater, which had been silent, erupted. Maybe they were laughing belatedly at the donut hole joke, but I think my aunt was the trigger.

Speaking of emotional contagion, "mirror neurons" are a hot topic in the field of neurobiology right now. So, they're subject to a fair amount of controversy. In brain-imaging studies featuring monkeys and humans, the observation of someone doing a task—like picking something up or eating a piece of food— causes neurons linked to task-associated muscles to activate *in the observer*. Observing someone else engaged in a task can cause someone to "mirror" that task vicariously in their mind. Mirror neurons might help humans pick up behaviors and techniques from others by pre-training our muscles while we observe, but it's a much larger leap to say that we also can mirror the mental and emotional states of others in this way. Could we use mirror neurons to replicate the feelings and behaviors of others? As of 2024, the scientific jury is still out.

Empathy is an important tool in the evolutionary arsenal of social animals that affect many of the principles of behavior. Within the frame of kin selection, empathy helps a rat better understand what is causing distress to its children or other relatives; the shared pain makes the rat work harder to alleviate it. Within the frame of cultural learning, the crow feels the visceral anger and fear of its flockmates as the caveman-masked human approaches, triggering an emotional contagion that helps convey the severity of the threat to the other crows. Within the frame of tit-for-tat cooperation, the chimpanzee sees how happy one of the older, respected males is when given a gift of meat. Maybe, if the chimpanzee can provide meat to that male, he will be aided in his political campaign for dominance of the group.

"I feel you" isn't just a metaphor. Empathy exists to amplify social and emotional bonds, to teach about threats and encourage preventative action, and to work as a kind of emotional information-gathering tool. All of these factors can help an animal and its offspring survive and reproduce. When an organism has both the more basic empathy and the more advanced Theory of Mind, that's when things start to get interesting, morally speaking.

Once Theory of Mind intermingles with empathy, brains interpret signals differently. When younger children watch someone else experiencing pain, brain activation is in the more fundamental areas associated with pain regulation and sensorimotor input. As people age, the registering of someone else's pain shifts to the prefrontal cortex, the area of the brain associated with abstract reasoning — thoughts focus on how the pain is affecting the other person, not just on the pain itself. What caused this pain? Who caused it? In *Mama's Last Hug*, de Waal brings up an anecdote from the Russian zoopsychologist Nadia Ladygina-Kohts, writing about a chimpanzee named Joni, back in 1935:

> If I pretend to be crying, close my eyes and weep, Joni immediately stops his play or any other activities ...

> He hastily runs around me, as if looking for the offender; looking at my face, he tenderly takes my chin in his palm, lightly touches my face with his finger, as though trying to understand what is happening, and turns around, clenching his toes into firm fists.

This passage is, as de Waal notes, an excellent example of empathic concern in a primate. The phrase "as if looking for the offender" is especially interesting. Joni both wants to comfort his keeper and find the source of her pain. Chimpanzees understand interpersonal violence. With older humans and chimps alike, seeing someone pushed to the ground is going to be much more emotionally charged than merely seeing them trip and fall on their own. The latter triggers feelings of sympathy and concern, while the former adds something more: a sense of injustice.

Once kids can distinguish intentional and unintentional harm, only "intentional" harm activates parts of the brain associated with reproach. Once the brain feels reproach, it looks to direct it at an offender. Sapolsky writes that children between the ages four and six feel injustice when they themselves are wronged. At the eight-to-ten-year mark, their sense of injustice extends outward. Their empathy broadens.

At last, the pieces are set for an animal to develop a nuanced sense of right and wrong. A bonobo trips over a log in the forest as her cousin watches. The cousin might feel sympathy, but nobody is to blame. Now, the bonobo is pushed to the ground by a bullying older sister. Is the sister to blame? The cousin thinks about how the victim felt upon being pushed, how this affected her mental state, and what the mental state of the sister might be. The cousin feels anger and reproach at the offender, and compassion for the victim, concluding that the push was "wrong."

Let's complicate the scenario. What if a human pushes a stranger out of the way of an incoming trolley. Is the pusher in the wrong? It's a no-brainer: of course not. The victim is still roughed up from being pushed, but the perpetrator's intentions were good; in fact, the perpetrator tried to save a life! Any human will say it was the "right" thing to do, morally speaking.

What if the stranger, who turned out to be large and heavy, was pushed in *front* of the trolley to save people further down the line? Now the push is morally ambiguous. By shoving the stranger in front of a trolley, the shover acted to save lives but sacrificed one in the process. Was this wrong? How does the family of the victim feel? What about the families of those rescued? Whether consciously or subconsciously, all these factors go into how we later judge moral acts.[23]

The more layers of complexity we add to morally ambiguous scenarios, the more perspectives we adopt and the more strongly we empathize. It gets harder and harder to make clear judgments of guilt or innocence. Humans want consistency. We want a set of laws to keep us from making these judgments over and over again. Unfortunately, nearly every story is unique. It's hard to judge each situation — with its own unique set of circumstances — by a single standard.

The older you are, the more complex your moral reasoning becomes. A man once filmed his toddler being presented the Trolley Problem with a toy train at the fork of two tracks — one leading to five action figures and one leading to a single action figure. The two-year-old solved this moral

---

[23] By the way, this is how we judge things *after* the fact. In the heat of the moment, we don't think, we either act or do nothing.

dilemma by taking the one action figure that was out of harm's way, putting it with the others, and then proceeded to hit all six with his train. "Uh oh!" The kid exclaims.

"Uh oh," indeed.

Lawrence Kohlberg is a foundational figure in the field of moral development who worked out his theory in the 1950s and '60s. Kohlberg split moral development into three stages: preconventional, conventional, and postconventional.[24] Let's use the Les Mis Dilemma (based on the Victor Hugo novel and subsequent musical) to illustrate each stage. Jean Valjean is poor and has a starving family to feed. No one will give him money or food. A baker has left a loaf of bread unattended. Should he steal it?

In preconventional reasoning, a person places the moral focus entirely on the consequences for the individual. "What's in it for Jean? How likely is he to get caught? Will feeding his family satisfy him?" Children usually use this level of self-interested reasoning. In conventional reasoning, the moral focus shifts to personal relationships and the law. For instance, you could say, "Jean *should* steal the loaf of bread because it is his duty as a husband and father to feed his family." You could also say, "Jean *shouldn't* steal the loaf because it's against the law to steal. What if everyone went around stealing bread when they were hungry?" Most teens and adults are here. Finally, in postconventional reasoning, you get into abstraction: "*No,* don't steal it. What of the baker's family? What if he is deeply in debt, or must sell this loaf to keep his home?" Alternatively: "*Yes,* steal the bread. It may be against the law, but who would

---

[24] There are many problems with Kohlberg's theory, but even today it serves as a useful model for thinking about the way our moral behavior becomes more complex.

value a single loaf over the lives of people? The lawmakers put Jean in this position in the first place." Postconventional reasoning is supposedly more rare.

It's difficult to tell if animals other than humans use Kohlbergian stages of moral reasoning. Most researchers would likely say that animals top out at preconventional reasoning — they are self-motivated, either for their own hides or their relatives', due to their evolutionary history. Our couch-destroying, guilty-looking dog from earlier probably fits the bill for preconventional reasoning. Evolutionarily "intentional" or not, maybe even non-human animals have ways of expanding their moral horizons.

When animals use both empathy and Theory of Mind to a sufficient degree, you can wind up with something astonishing: the emergence of "truly" altruistic behavior. In kin selection, an animal will help another to protect its genes, which it shares with extended family members. But what happens when an animal works to protect — and even puts itself at risk for — the survival of a complete stranger, maybe not even a member of its own *species?*

12. Empathy Overflowing

"Sympathy beyond the confines of man, that is, humanity to the lower animals, seems to be one of the latest moral acquisitions."
— Charles Darwin, *The Descent of Man.*

They've lost Annie. She dies at ten years old on April 23rd, 1851. There is an agony in the room that defies description. Charles is there at the doctor's in Great Malvern. Emma is in the late stages of pregnancy and could not make the journey. Darwin now has the impossible task of delivering the news to her. He feels physically ill.

They have lost a child before– their third child, Mary, lived for three weeks before passing. This is not an uncommon occurrence in their part of the world and time period. Annie's death is affecting Charles differently– he has gotten to see her grow up, to know her as a cheerful, loving, and affectionate daughter. Now he has seen the effects of disease– possibly tuberculosis– take all of that away. He writes her memorial while wrought with grief, filling the paragraphs with vivid descriptions of her kindness and beauty.

"Besides her joyousness thus tempered, she was in her manners remarkably cordial, frank, open, straightforward natural and without any shade of reserve. Her whole mind was pure & transparent. One felt one knew her thoroughly & could trust her: I always thought, that come what might, we should have had in our old age, at least one loving soul, which nothing could have changed."

And yet everything has changed. Seeing his daughter fall ill, doing his best to get treatment for her, and finding that treatment still come up short– it is more than Charles can bear.

"We have lost the joy of the Household, and the solace of our old age:— she must have known how we loved her; oh that she could now know how deeply, how tenderly we do still & shall ever love her dear joyous face."

Many people will tell you that developing evolutionary theory undermined Charles' Christian faith. They are wrong. It was Annie. Her death diminished his religiosity more than any scientific treatise or discovery could have. Afterward, he is far more willing to share his spiritual doubts with colleagues.

Many religious people experience a strengthening of their faith through adversity, but this does not happen to Darwin. After this tragedy, Darwin's analytical mind struggles and fails to reconcile a kind and loving God with a universe that would kill his little daughter from such a terrible disease. Every day is a struggle. Some people claim that Darwin is so overwhelmed with sorrow that he cannot attend Annie's burial.

But his faith is not gone entirely, and neither is his will. The Origin of Species will not be published for another eight years. His family and his research provide some small measure of solace. Yes, there are larger implications to his research on evolution, but they are not atheistic.

God has been misplaced in Darwin's world, not lost. In the midst of the worst grief imaginable, the Darwins are not alone. There are others grieving– for the Darwins' loss, for their losses, for both. They mourn with them. They feel their pain. They wish them well. There is a kind of beauty in it, the collective outpouring of empathy that humans exhibit toward one another in the face of tragedy. It overflows from our beings– and it must, Darwin thinks to himself, exist for a reason.

There must be some higher order, some meaning to all of this, all of the joy and suffering that Darwin sees in the world. If God was not at his daughter's bedside, then where? Where could He be hiding?

///

Empathy and sacrifice are not unique to our species. Dolphins are among the most intelligent species on the planet, with a brain-to-body size ratio beaten only by humans.[25] They use a complex system of communication consisting of clicks and squeaks that we've been trying to decipher for decades. They play with one another purely for the fun of it. They give each other names. They use mud as a tool to disorient and trap fish in coordinated hunts. They are also, like their larger whale cousins, a group of species that rescues humans for seemingly no reason other than goodwill.

In the book *Dolphin Confidential: Confessions of a Field Biologist,* a group of marine biologists were following a bottlenose dolphin pod when one dolphin suddenly broke off, heading into deeper water. This was unusual behavior; they do not usually leave foraging sites so abruptly. The team kept pace in a motorboat, following the lone dolphin and later, its entire pod, which joined in leading the researchers. Minutes later, the dolphins "stopped, forming a sort of ring around a dark object in the water."

---

[25] Brain-to-body size ratio is not the perfect metric for cognitive ability, but it works pretty well with larger animals.

That dark object turned out to be the floating, hypothermic body of a teenage girl who had attempted suicide, leaving a note tied to herself. Thankfully, the researchers were accompanied by people with lifeguard training, who were able to save the girl's life. The author of *Dolphin Confidential*, a scientist and science journalist named Maddalena Bearzi, says, "If we hadn't found her, if the dolphins hadn't led us offshore when they did, to that specific place, she would have died." Did they detect the girl through echolocation, and did the first dolphin recognize that the human researchers could help? There must be an explanation for the dolphins' actions. Was it pure randomness, or an act of deliberate empathy?

This is not an isolated incident among cetaceans, the group that includes whales and dolphins. In one report, a dolphin pushed a drowning boy, who had fallen overboard, to the surface, allowing him to climb back onto his boat. In another, dolphins surrounded a diver in the midst of a group of sharks, protecting him and driving the sharks away. In a viral video by *BBC Earth,* veteran whale biologist Nan Hauser recounts a massive humpback whale swimming up to her, trying to push her away from their location and out of the water. Hauser didn't understand why the whale was doing so until she noticed the massive tiger shark swimming nearby — the whale seemed to be trying to protect her from it, shielding Hauser from the shark's view.

There are hundreds of similar reports, many verified by scientists and discussed in scientific journals. Researchers confirm that humpback whales frequently protect seals, and even sunfish, from attacks by orcas, with seemingly no benefit to themselves. In fact, a paper in the journal *Marine Mammal Science* points out that humpback whales actively put themselves in *danger* when attempting to protect other

Above: Cetaceans like whales and dolphins will go out of their way to help other species. Top photo by Louis Herman. Bottom photo by NASA. Accessed via Wikimedia Commons.

animals. It's difficult to "logic away" these altruistic behaviors using evolutionary theory. Kin selection doesn't make sense in this context; humans and sunfish are many millions of years removed, evolutionarily, from dolphins and whales. It wouldn't appear that the cetaceans expected a reward. Dolphins and whales often swam off immediately after seeing their human or dog or seal or fish to safety. What kind of reward would a whale expect from a fish? The source of this altruism remains unresolved.

It's important to note that *some* accounts of whales and dolphins intentionally saving people don't meet the "disprovability" threshold that is essential for science. We can't be sure what led the animals to action. Animal motivation, even with a well-designed scientific experiment, is tough to prove. That's especially true for large-brained cetaceans. Accepting those limitations, let's speculate.

Empathy's a powerful tool for animals, both as a teacher and motivator. Given what we know about whales, with their communal learning, enjoyment of play, and complex social structures, we can be pretty sure that these animals rely heavily on empathetic brains. We've discussed the benefits of empathy for kin selection, and cetaceans do make great efforts to aid and protect their own kin. But why extend this behavior to other animals?

I believe that empathy doesn't have to be tailored to protecting one's own kin to be effective. Natural selection could favor empathy without necessarily selecting for individuals who were *only* empathetic to close relatives. Kin selection isn't perfect, and it doesn't have to be. Cetaceans aren't close relatives, but it's conceivable that when they see humans, dogs, or even sunfish in danger, it may trigger a strong empathic response in their brains. After all, have you ever watched a nature documentary and felt bad for the poor, cute, defenseless baby gazelle in the jaws of a hungry lion? Or the lioness who can't provide enough milk for her cubs after too many failed hunts? Or the first 10 minutes of Bambie? We empathize with animals that aren't close relatives all the time. We've even built a specific website to let us know if a dog dies in a movie, so that we can avoid watching it and ruining our day.[26]

We feel that way about other animals because they activate the empathetic responses hardwired into our brains — we feel their pain. Whales and dolphins may feel the same way when they see a human in danger of a shark attack. Seeing humans in pain causes feelings of discomfort in the cetaceans, motivating them to relieve that discomfort. A recent study found that elephant brains (that is, the brains of elephants that haven't experienced poaching) react to seeing humans in much the same way that human brains react to seeing dogs. In other words, they find us cute. Maybe whales find us cute, too. It might explain their motives. How could a giant elephant or whale say no to a cute bipedal primate in distress?

Charles Darwin himself was an unusually empathetic person. Maybe that's why he so strongly emphasized the importance of what he called "sympathy" in his theory of moral development, as we discussed earlier. Once an avid hunter, Darwin gave up shooting birds entirely when he saw one of them writhe in pain after being hit by a shell. He was also a strong early advocate for animal welfare, calling for sympathy for animals as "our fellow brethren in pain, disease, and suffering." To witness suffering distressed Darwin. He could not stomach watching a young child undergo surgery during his two years at medical school. The memory of seeing slaves

---

[26] If you're interested, it's www.doesthedogdie.com

tortured in Brazil haunted him for years. Darwin was a strongly empathetic human. Many of us are.

Empathy causes some animal behavior to go above and beyond what the pure logic of natural selection theory deems "advantageous." Truly altruistic behavior — like a whale saving a human — results from a sort of "empathetic spillover." The primatologist Christopher Boehm writes that "somehow, some of this generosity is simply 'spilling over' owing to natural selection's inefficiency ... overall the individual advantages of nepotistic generosity are so strong that moderate amounts of costly extra extrafamilial generosity could be 'piggybacking' on them genetically." Even though nature didn't select for the whales' inter-species altruistic behavior itself, it selected for the temperament that *leads* to these kinds of behaviors. Altruism may emerge from animals making a mistake in kin-recognition. Rather than use the term "mistake," though, I prefer painter and TV host Bob Ross's phrase: "a happy little accident."

We *should* be happy about it. Empathetic overflow lets animals extend their feelings beyond kin. This has massive implications for human morality: not all moral behavior must be explained as adaptive. Kindness, caring, and sacrifice don't need to be the byproduct of a cold, Machiavellian calculation aimed at spreading genes. Moral behavior is not necessarily "useful" to us or our genetic progeny, because the innate tendency toward this kind of behavior could still be selected for through helping one's relatives. This evidence supports Darwin's idea that moral behavior evolved naturally. Truly altruistic behavior, arising as a by-product of "selfish" behavior — that would be a happy little accident, indeed.

Before we start seeing Mother Nature through rose-tinted swim goggles, a few facts need clarification. First, a word on dolphins: they are not always kind. In addition to rescuing dogs, dolphins reportedly brutalize porpoises.[27] Researchers Harry Ross and Ben Wilson found that 63% of dead porpoises found stranded around Moray Firth in Scotland had "died from trauma characterized by multiple skeletal fractures and damaged internal organs." They were also able to attribute tooth marks on the corpses to the bottlenose dolphins and observed several instances of dolphins attacking porpoises.

These "crime scenes" were observed by people with cameras, who filmed gangs of dolphins beating porpoises to death. One team of researchers described it as "perhaps the worst example of inter-specific aggression any of us had ever seen. [A] young female had literally had the life beaten out of her." In another video, a dolphin kills a baby porpoise and plays with its corpse. Lots of animals kill, but researchers have a hard time understanding why dolphins would go out of their way to kill other cetaceans. In some instances, they seem to be killing for sport, which is a behavior exhibited by very few animals, humans among them. There is a Yin and Yang to moral development: as our capacity for kindness increases, so does our capacity for cruelty.

Male dolphins can also commit infanticide, presumably to send females back into estrus, so that the male can mate and sire his own offspring. This behavior is shared with lions and polar bears, but for an animal as cognitively advanced as a dolphin, it may seem surprising. Do other dolphins recognize this behavior as "wrong"?[28] Perhaps, to dolphins, it isn't "wrong" at all, but rather an ancestral behavior encoded in their

---

[27] Using these terms can be problematic, because we are attributing human acts to non-human animals. Some researchers believe dolphins are intelligent enough for these terms to apply,though.

[28] More on moral relativity later.

genes. In any case, even playful, "lighthearted" animals like dolphins are capable of horrific behavior by most human standards. It's an important lesson to keep in mind before we glorify an animal and wish that humans behaved the same way. Kindness and cruelty go hand-in-hand, flipper-in-flipper.

Another caveat — just because animals go out of their way *in some cases* to save someone outside their own group doesn't mean they won't discriminate. In lab experiments, animals almost always prioritize helping those they identify as "kin." Mice are more likely to help mice from the group they grew up in. Capuchin monkeys are more likely to share food with members of their own clan. Animals can act altruistically, but they still maintain basic evolutionary common sense: if you must choose, protect your own genes first.

Humans are also heavily impacted by ingroup-outgroup discrimination. It's true, we work harder to help our own. And yet, in spite of our self-interestedness, humans may be the most empathetic species on Earth. Human empathy regularly extends far beyond familial relationships to entire countries, even for the planet as a whole. Watching animals fleeing a logged forest evokes discomfort in people. We wring our hands at news of an oil tanker spilling into the ocean, but the spill has no immediate impact on most of us. Dawn, the detergent company, launched a series of ads following a B.P. oil spill in 2010, showing their detergent being used to clean ducks and other animals affected by the spill. Whether the detergent was helpful in the long run is an open question, but the ads tugged at our heartstrings. The company wanted to give us that warm fuzzy feeling– along with a subliminal desire to buy Dawn products. The ducklings remain on Dawn bottles to this day.

Perhaps these ads move some of us because of their larger implications for our own long-term survival, but most of us are focused on the ducks. Human brains are hardwired to fear immediate threats, not long term threats, even when the long term threats are far more deadly. Plenty of movies feature victims escaping the Jaws of a murderous shark, but few where protagonists must beat the deadly monster of malaria: each year eight people die from shark attacks, half a million from malaria. Even though malaria is a much bigger danger to us in the long term, we're more concerned with immediate threats like a shark, a monster, or quicksand. This brings us back to the oil spill: the reason we find the plight of animals emotionally compelling is because they invoke immediate feelings of empathy — we identify with the suffering creatures.

Consider the giant panda. As far as ecologically important species go, pandas are not exactly priority number one. They're large mammals that subsist entirely on bamboo — not the most nutritious food source. Pandas are so reluctant to breed, trainers must make a huge effort to get them in the mood, and there are celebrations and high-fives all around when a pair finally manages to get it on. Sure, the panda might help spread some bamboo seeds around and graze down the vegetation here and there, but it's not a keystone species. A single species of coral has a much broader impact on its ecosystem — coral foster dozens of reef species that rely on them for survival, and they are many millions of years old.

The mascot of the World Wildlife Fund is not coral. It's the giant panda, and that panda brings in millions and millions of dollars for wildlife protection every year. Giant pandas bring in the big bucks for one main reason: they're adorable. Big eyes, fluffy, seemingly so docile ... It's heartwarming to see these fuzzy bears waddle around. They spark the same empathetic, protective instinct in us as big-eyed, helpless human babies do, activating the same "empathetic triggers" in our brains. Coral can be nice to look at, but it's not what most people would describe as cute. Most people probably don't even consider coral as animals.

This is an example of a larger phenomenon that folks in conservation refer to as "Charismatic Megafauna." Larger, cuter animals get more attention than smaller, more ecologically important conservation targets, simply because we empathize more with them. The human brain's evolutionary bias toward appealing appearances factors into multi-million-dollar conservation decisions. We shouldn't blame ourselves for this bias. It's simply a bias. We need to accept it, and to be aware of it when we make decisions about which species to prioritize protecting.

The Giant Panda has become a face for animal conservation. Photo by J. Patrick Fisher. Accessed via Wikimedia Commons.

Humans are now so successful in their drive to be fruitful and multiply that they directly or indirectly threaten a huge percentage of the world's species. This wouldn't bother other species in the slightest. An invasive python would be happy to go around eating all of the other animals, even if those animals' populations started to decline from rapid overhunting and eventually crashed the python population as well. A swarm of army ants would never say to itself, "Hey, do you think we're getting overpopulated? This is probably having a detrimental effect on the rest of the ecosystem as we strip the rainforests bare to build our colonies." Humans are the exception — we are cognitively advanced enough to consider the threat of extinction, to understand what we are doing to other species. We have our empathetic overflow to thank for that. We don't want the pandas to die. We don't want the cute dolphins to go extinct. We often have a hard time recognizing the long-term threat that climate change and extinction poses, but our empathy for other animals just might be our saving grace.

To top it all off, even though showing empathy for creatures other than our close relatives could just be a happy little accident, it has real consequences for our survival. Without this "accident," we might not have any chance at all to tackle the extinction crisis. Our overabundance of empathy– our love for our physical and genetic neighbors– may in the long run turn out to be a pre-adaptation that gives us the emotional motivation to save the planet, and with it, ourselves.

13. What Darwin Got Wrong

> "How fleeting are the wishes and efforts of man! How short his time! Consequently how poor will his products be, compared with those accumulated by nature during whole geological periods. Can we wonder, then, that nature's productions should be far "truer" in character than man's productions; that they should be infinitely better adapted to the most complex conditions of life, and should plainly bear the stamp of far higher workmanship?"
>
> — Charles Darwin, *The Origin of Species*

There is a human-shaped elephant in the biological room. The Origin of Species has skirted around the topic of human evolution like a man with paper shoes around a mud puddle. Nonetheless, the need for humans to make everything about themselves is incessant– of course people are applying what Darwin has written about natural selection (along with Wallace's work) to their own origins. Even in the few years since its publication, this has started to generate a slew of new ideas, some intriguing, others terrifying, about the human species. Campus coffeehouses and pubs light up with revolutionary thought, some of it insidious in nature. The political climate in 1871 is changing. It is time for Darwin to finally throw his own bowler hat into the ring.

The Descent of Man covers head on what the Origin of Species avoided altogether. An emboldened Darwin frames human evolution from his natural selection standpoint. One of the major topics discussed is the origin of human morality– it is, in some ways, where Darwin has at long last found a place for his missing God.

Empathy and sacrifice– Darwin has gone through a lot while searching for the origin of these human characteristics. It is, perhaps, only after a lifetime of experiences around the world and at home that Darwin can see: the source of our morality is both external and internal. Our moral behavior comes from our evolutionary history and is both reinforced and made relevant by pressures from the world around us. He argues that the pressures to create morality are seared into the universe by a Creator– just like gravity. He also argues that humans descend from an ape ancestor. The magazines love depicting his bearded, balding head attached to a chimpanzee-like body.

Darwin dedicates three whole chapters in his book to describing where morality– that hallmark of what makes humanity human, comes from. Creatures that care for one another fare better in a harsh world than those that do not. Social systems of right and wrong win out over "dog eat dog." The Golden Rule flourishes from countless generations of strife– a powerful byproduct of natural selection. Here at last is a Golden Rule that makes sense to Darwin. It is a rule that emerges from a universe governed by laws.

Eureka. Countless pages of notebooks whirl about in his writing frenzy. Here is how a benevolent God can exist alongside a hundred thousand parasitoid wasps. Here is why God did not intervene to save his daughter's life. God's plan– if there indeed is one– was set in motion at the beginning of time. This great, noble, cosmic laboratory experiment has been running on its own ever since, and humanity is simply one of its end results. Finally, a God– and a universe– that his grieving scientist mind can understand.

There are many ideas Darwin puts forward in the Descent of Man that are flawed. He is wrong on many accounts, not least of which are his arguments that Europeans are more highly evolved, or that men are far more intelligent and inventive than women. Despite being progressive for his time, he is still a wealthy Victorian and this cognitive bias influences his views on human evolution. He puts forward that evolution progresses from "worse" to "better," a concept outdated in 21st century biology. He has not worked out the precise mechanisms of how traits are inherited. The actual speed and timing of human evolution is unclear. In spite of those biases and errors, in spite of the looming gaps in Darwin's knowledge, his description of human moral evolution is world-changing and still relevant to this day. It is a signal of things to come. It is a new mirror for the human species, one that might finally let us see ourselves clearly in the reflection.

Charles Darwin in 1869. Photo by Julia Margaret Cameron in the public domain. Accessed via Wikimedia Commons.

///

Some readers may find the prior chapter on empathy too optimistic. You could easily say that humans are incapable of making changes fast enough to keep pace with the world they created — it's just a matter of time before environmental change gets completely out of hand. To that I say: we've done it before. Humans are an astonishingly adaptable species, capable of great change in no time at all. We're not alone in this regard — plenty of species can alter their form and functions in the blink of a geological eye. The potential for that sort of rapid change brings us all the way back to Darwin, and perhaps one of the biggest ideas about evolution that he got *wrong*.

Doctors, scientists, and race-car drivers alike will always be limited by the technology of their day. No matter how good a racer you are, you won't win the Indy 500 in the 2020s with a car from the 1920s. No matter how good a doctor you are, you can't successfully perform brain surgery using only a rock. No matter how good a scientist you are, you can work only with the information that the world around you renders observable.

That said, scientists and natural philosophers have a historically good track record when it comes to making the most out of the available resources. The most famous case of making the discovery of a lifetime out of practically nothing is probably Eratosthenes of Cyrene, a Greek philosopher and mathematician who lived sometime around 300 BCE. Using two sticks, a sundial, basic mathematics, and a guy he paid to walk around for a couple of days, Eratosthenes devised an experiment that not only proved the Earth was round, but also measured its circumference.

The city of Cyrene was located close enough to the equator that, during the summer solstice, the sun peaked directly overhead. Its rays were so direct that they shone into a Cyrenian well and illuminated the water at the bottom without touching the well's sides. First, Eratosthenes set up a pole in Cyrene and measured the length of the shadow it cast at noon. Then he recorded the same measurement in Alexandria. Sure enough, the Alexandrian shadow cast a greater angle. That difference could only be possible if the sun were shining onto a curved surface, which would explain the difference in shadow angles between two distinct points on that surface. Eratosthenes also knew the approximate distance from Alexandria to Cyrene, and from this he could calculate the circumference of the *entire Earth.* By his calculations, the circumference of the planet was roughly 25,000 miles (40,200 kilometers). In reality, the Earth has a circumference of 24,901 miles (or 40,075 kilometers) around the equator. To quote the legendary scientist and science communicator Carl Sagan, "Pretty good figuring for 2,200 years ago." Eratosthenes was born almost two millennia before satellites or airplanes. He couldn't physically see the round Earth in its entirety, but he was able to infer its shape from what he *could* see.

Likewise, Charles Darwin developed his theory of natural selection's influence on physical traits without fully understanding *how* those traits were passed on. Like Eratosthenes, Darwin had to infer from what he *could* observe, rather than observing evolution directly. The structure and function of DNA was unknown until nearly one hundred years later, and yet Darwin made predictions about selection, evolution, and the course of life's history all the same. Because Darwin did not fully understand *how* traits get passed on, he had to make some pretty big assumptions. Those assumptions were a big reason why Darwin got some crucial ideas wrong in his evolutionary timescale: specifically, he had it working too *slow*!

It's rare when the popularizer of a theory actually underestimates its strength. Darwin believed that evolution moved so slowly, acted so incrementally, that there was no way that we could observe it directly. Gradualism posed a problem to the falsifiability of evolution — meaning "the possibility of proving it wrong." Without knowledge of mutation, DNA, and epigenetics,[29] Darwin would have a hard time understanding that real, observable evolution happens not just across millions of years, but sometimes in the span of decades as well. We now know that, under the right circumstances, evolution can move at a breakneck pace, splitting species, merging species, and leading to wild adaptations. Out of all of the advances made in our understanding of evolutionary biology in the century and a half since Darwin, our direct observation of evolution might be the most important of all.

To be fair to Darwin, the term "evolution" did not appear in the first edition of *The Origin of Species*. It was added to a later version in 1872. Instead, Darwin wrote of "natural selection" and "descent with modification." Today, "evolution" is understood as the descent with modification of different populations and species, and natural selection is just one mechanism through which evolution takes place.[30] Even on its

---

[29] Epigenetics refers to genes' ability to tweak themselves during an organism's lifetime.

[30] Others include genetic drift (a rare gene becomes common in a small population), sexual selection (choice of mates drives new traits), and hybridization.

own, natural selection is capable of producing rapid evolution — so long as the pressure to adapt is strong enough.

Sometime in the 1970s, a biologist named Tom Schoener decided to introduce a bunch of *Anolis* lizards, all from the same populations, to some tiny, remote, lizard-less islands in the Bahamas. Each of these islands had different vegetation. Some had plants with broad leaves and thick branches, others had plants with shorter leaves and thinner branches. Roughly a decade later, scientist and science writer Jonathan Losos went out and caught the descendants of the lizards Schoener had introduced, measuring the lengths of their legs. He found a fascinating result: the lizards living on islands with thick branches had significantly longer legs, while the lizards on islands with thin branches had evolved substantially shorter ones. Natural selection was running its course.

Repeatability is critical for science. Losos and his research team repeated this experiment with a different anole population, different islands, but a similar spread of vegetation. Sure enough, they got the same result: longer limbs on the islands with big plants, shorter ones on the islands with small plants. It's a small but powerful experiment, proof of just how quickly natural selection can start to drive organisms to change. Today, lots of astounding research continues with anole lizards — most recently, teams of scientists looked at how hurricanes changed the bodies of anole populations, lengthening their limbs and strengthening their grip. Hang in there, lizards.

Scientists are even watching speciation occur in the finches named after Darwin himself. Jonathan Weiner's Pulitzer Prize-winning book, *The Beak of the Finch*, tells the story of Peter and Rosemary Grant, a pair of scientists who spent decades in the Galapagos tracking populations of Darwin's finches. They had captured, marked, and measured just about every finch on those islands during their many years of research, recording their lineages as entire finch families lived, grew, and died out.

1. Geospiza magnirostris.
2. Geospiza fortis.
3. Geospiza parvula.
4. Certhidea olivacea.

Above: multiple varieties of Darwin's Finches with their prominent beak differences. Drawing by John Gould in the public domain.

The Grants' work changed our view of evolution. In one study on the island of Daphne Major, they discovered that the smaller-beaked *Geospiza scandens* were getting closer and closer in average beak dimensions to the larger-beaked *Geospiza fortis*. More testing and observations revealed the reason: the birds had begun to hybridize. Not only were more and more birds hybridizing each year, but those hybrids had better success passing on their genes than the "purebloods" of either species. Whatever advantages hybridization gave those birds, it was working. Natural selection could eventually mix the two species on the island so thoroughly that, under the right circumstances, they converge, creating a brand-new species. Darwin would have been wowed by the sheer speed at which his finches are evolving.

On another Galapagos island, an invasive parasitic fly has been wreaking havoc on finch populations, parasitizing nestlings and leading to a nearly 90% mortality rate in the Medium Tree Finch. The species has been on the verge of extinction. In the face of being wiped out, the Medium Tree Finches have developed a way to save their genetic lineage. Katharina Peters and her research team made an intriguing discovery: the finches have begun to hybridize with Small Tree Finches, another species on the island that's likewise hammered by nest parasitism. Incredibly, Peters found that the hybrids of these two species are more resilient to the parasites than either species is on its own. Tree Finches are able to give parasite resistance to their offspring by combining their genetic material, giving their species a chance at survival in a new form. The pressure and comparative advantage toward hybridization is so strong that, in a couple of decades, another species of finch will likely soar above the choppy seas of the Galapagos.

We've known for a long time that mixing genes protects against disease. In fact, that very concept is the most popular theory for the evolution of sex. By combining our DNA with a partner's, we trade half of our own genetic material for the advantage conferred by merging it with someone else's — we battle an unyielding enemy. The most rapidly evolving organisms around: pathogens.

Lizards and birds are fun to live alongside. Diseases are not. Unfortunately, the latter is probably the most easily observable case of rapid evolution out there. For example, when antibiotics were first invented, they were extremely effective for just about any bacterial disease. Now, not so much.

According to the CDC, 18 of the most prevalent bacterial and fungal diseases now resist antibiotics. There are probably plenty more out there. Where penicillin used to be *the* treatment for just about any bacterial disease, antibiotic-resistant bacteria now infect around three million people a year and kill over thirty-five thousand in the U.S. alone — hundreds of thousands more are killed by antibiotic-resistant organisms worldwide.

All these statistics tell us that in our race to combat disease, humans have caused the rate of pathogen evolution to go absolutely wild. Bacteria and viruses mutate quickly, altering their cellular structures to resist the latest chemical compounds and vaccines we manufacture. Even now, new variants of the Coronavirus continue to spring up and plague our species. The arms race between humans and diseases will never end, at least not in the foreseeable future.

Rapid evolution of resistance doesn't stop at the microscopic level, either. After decades of spraying our crops with pesticides and genetically engineering corn to kill unwanted bugs, many insects now carry a mutation that protects against pesticides. As the mutant bugs survive and reproduce, the genes for resistance will soon spread. Some unwanted plants no longer die from weed killer the way they used to. Human activity has become a major driving force for natural selection. Sure, we can develop new antibiotics and pesticides and weed killers, but any fight against nature is one we will surely lose. Like Darwin, we have continued to underestimate the sheer speed of evolution.

Luckily, humans have a trick up our sleeve: it's not just birds and bacteria that can evolve fast. We, too, are capable of great and rapid change, but instead of our DNA shifting, it's something a bit less tangible. We possess an adaptable cultural and moral sense, one that can change at the sight of a God, fall of a tree, or drop of a bomb. Just like the finches in the Galapagos or the lizards in the Bahamas, our adaptability has allowed us to weather the "slings and arrows" of life.

Unfortunately, not every organism is well suited to the environmental changes that humans cause in the quest to accommodate our needs. Nor are all organisms capable of rapid adaptive change in the first place — it depends on the context. As much as humans adapt to the world around them, the world has to adapt to us, too. The results are mixed, with clear winners and losers: those that can change to live with human changes and those that can't. We've seen how species can hybridize in the face of adversity, but what happens when we cause entire worlds to collide?

14. Trials, Tribulations, and Unintentional Invasions

> "But Natural Selection, as we shall hereafter see, is a power incessantly ready for action, and is immeasurably superior to man's feeble efforts, as the works of Nature are to those of Art."
>
> — Charles Darwin, *The Origin of Species*

Just outside my Chicago apartment window, she perches on a nearby branch, which bends ever so slightly under her miniscule weight. She cocks her tan and brown head and lets out a polite call, far less aggressive than an acorn woodpecker's: "Chirrup? Chweep! Chweep! Chirrup?"

She is a house sparrow, a longtime urban colonist and a wildly successful one at that. A couple hundred years ago, her great, great, great, great ... Her *very* great grandparents came here from Europe, likely on a fleet of cargo ships. Some of her ancestors were deliberately introduced to the Americas. Others came by accident. Wherever the colonists colonized, house sparrows followed in their wake, thriving in the farmland and towns that the Europeans so eagerly terraformed to resemble their homelands. Now the house sparrow stretches far and wide, having successfully established a foothold on every human-populated continent on Earth.

The history of house sparrows and humans goes a long way back, all the way back to the beginnings of civilization in the Middle East. When humans first began practicing agriculture, house sparrows took well to the comfortable, seed-filled fields that people created. Centuries passed, and slowly at first, but then with terrifying momentum, the human territorial range began to expand. More civilizations sprung up out of the Middle Eastern cradle. The birds followed. Cities and towns grew large enough to host burgeoning human populations, presenting more opportunity for the birds. One species' waste is often another species' food, and the sparrows were no exception. There's even evidence that house sparrows evolved to digest the starch from our foods, similar to human populations' evolving to digest the lactose in milk after infancy.

Natural selection played a role in altering the metropolitan house sparrow's behavior. The ecosystems we engineer in farmlands and cities alike host a set of conditions very different from any naturalized forest or grassland. Away from humans, it often pays to be careful with one's food choices. Many plants and animals, after all, are poisonous, and can harm or even kill you if you eat one by accident. It's good to be cautious around other animals as well. It takes just one slight misjudgment of another creature's intentions to wind up as its food.

The house sparrows we see living around us are true daredevils. They're not picky about what they put in their beaks and are willing to try just about anything we leave lying around that looks remotely edible. They're not afraid to get up close and personal either; they are far less timid around people than most birds. I have seen house sparrows digging at Chinese food from a discarded take-out box. I have seen them hang out under outdoor restaurant tables for scraps , waiting until the last possible moment to dodge any incoming feet. At a cafe in Spain, I witnessed one house sparrow fly right up onto my plate, grab a tiny chunk of uneaten French fry, and fly just a few feet away to scarf it down, almost taunting me. As the bird polished off a meal I had spent far too many euros on, I had to stop and admire the beauty of nature.

Above: a house sparrow perches on a branch. Photo by Imran Shah. Accessed via Wikimedia Commons.

House sparrows get away with their petty crimes. They even commit larger ones, like displacing native birds. Few urban animals pose a threat to house sparrows, except maybe a housecat or some unusually determined people, and furthermore, there aren't a whole lot of substances in urban environments that can kill them. Humans try to keep poisonous plants and insects far away. Over generations, birds that were more willing to get close to humans and try the food they left around would eat better. That extra food fed more offspring, passing on daredevil tendencies— via genetics, cultural conditioning, or both— to a large number of offspring. We would expect to see more boldness and less neophobia (fear of new things) in sparrows that live in cities, and that is exactly what scientists are finding.

Thanks to natural selection, house sparrows weren't just well suited for living among humans, they were *pre-adapted* to settling anywhere we went. Even today, as industrialization and development make their way into the heart of the Amazon rainforest, the sparrows follow, using our paved highways to guide them from town to town, city to city. From humble beginnings in the Middle East, *Passer domesticus* has followed agriculturalists to the ends of the Earth.

Not all bird species adapt to life among humans quite so well. Different species can have wildly different evolutionary histories. Humans began practicing agriculture in fertile Middle Eastern river valleys, exactly where house sparrows evolved. We convert wilderness into farmland by making it more like those river valleys — flat and open. It's only natural that house sparrows would take to land resembling their ancestral habitat.

Many species evolved where humans were scarce, like the middle of the Amazon rainforest or the Galapagos Islands. African mammals spent enough time evolving around humans that they recognize us as a threat and know to keep their distance. When Darwin first landed on the Galapagos, he noted that he could walk right up to the native birds and practically grab them in his hands.[31]

Complete incompatibility with human civilization is a vulnerability for animals. Because of the extent of human development, tolerance to humans has become highly useful for survival. The house sparrow now outcompetes many

---

[31] A lot of predator-avoidance behavior is learned. Galapagos finches have no natural predators, so there would be no reason to be afraid of *any* animal, regardless of what it looked like.

songbird species for nesting space and food resources. House sparrows are displacing other bird populations by outperforming in human-altered landscapes all over the world. Other species are fighting an uphill battle.

Countless invasive species outcompete and displace native ones, either because invaders lack predators in their non-native environments, deal with humans better, resist certain diseases, or for a slew of other reasons. We often talk about climate change like it's the only driving force in the present-day mass extinction, but the introduction of invasive species through human activity may in fact be doing just as much damage. If we think that preserving the world's most threatened species is the moral thing to do, then it's imperative that we take action to control invaders before they wipe their native competition out of existence.

But what can we actually do?

15. The Men Who Shoot at Goats

"Certainly, no fact in the long history of the world is so startling as the wide and repeated exterminations of its inhabitants."

— Charles Darwin, *The Voyage of the Beagle*

There are a few common tactics used to control invasive species, and none of them are perfect. One tactic involves introducing the invader's common predators. The predator theoretically controls the invader's population, giving the native competitors a fighting chance. Of course, this "solution" can quickly turn into an "old lady who swallowed a fly" story: the predator species *introduced* to stop an invader itself becomes widespread and invasive. Ecology is far from an exact science. We can't be certain exactly what effects a new species will have on the ecosystem, even when we make well-educated guesses.

The humble cane toad is a famous case of biological controls gone wrong. Looking to get rid of the cane beetles ravaging their sugarcane crops, farmers introduced one hundred cane toads from Puerto Rico to Australia in 1935. There are now nearly *two billion*.

Cane toads pose a major threat to specific Australian predators like snakes, which can die from eating the *extremely* toxic toads. They're a menace to the ecosystem in general, since they indirectly harm native plants by eating the insect-eating animals — without this control on the insect population, native plants get devoured. Introduced as a biological control, cane toads are now one of the world's most notorious invasive species, a self-made Plague of Frogs.

Above: The cane toad, once envisioned as a biological control, has gotten completely out-of-control in Australia.

Another anti-invasive strategy is the introduction of a disease or chemical custom-designed to bring down an invader's numbers. Australia's government used bioweapons for its European rabbit problem. With its numbers rising to around 10 billion in the mid-1990s, the invasive rabbit species devastated plant life and caused a significant ecological chain reaction in the process. After many failed attempts to control the rabbit population — poison, fencing, and some milder diseases — scientists introduced the rabbit haemorrhagic disease virus, or RHDV. This one did the trick, wiping out a massive percentage of the rabbit population. Soon native mammal species began to increase in number all across the country. Populations of invasive cats and foxes, which sustained themselves largely on the invasive rabbits, also began to crash. A major invasive problem had been dealt with, seemingly in a single blow.

Diseases can wipe out organisms terrifyingly fast. That said, there is often a time limit to their effectiveness. Even in the face of devastating plagues, life finds a way and develops immunity. With ten billion rabbits, even if a tiny 0.000001% of the rabbit population resists the disease, that still leaves ten-thousand individuals alive to breed. And rabbits breed like, well, rabbits.[32] Without a constantly updated virus, it won't take long for that rabbit population to return to its unwieldy numbers.

When biological and chemical controls fail, there's always brute force. Sometimes we simply take out invaders the old-fashioned way. This might involve wide-scale weed-pulling, assigning hunting bounties to certain species, or in the case of goats on the Galapagos, assembling a strike force of helicopters and men with sniper rifles.

You read that right. Domestic goats managed to find their way onto the very same Galapagos Islands that Darwin himself visited. There were three goats in 1959, then forty thousand in the 1970s, and then hundreds of thousands by the '90s. This posed a major problem for the fragile ecosystems of the Galapagos which, with goats chewing up all the native plant life, were in danger of some serious population crashes. In 1997, a multi-organizational initiative, whose members included the

[32] Rabbits have a month-long gestation period, give birth to up to twelve babies in a litter, and can get pregnant the day after giving birth.

Charles Darwin Foundation, started: "Project Isabela." It was a top secret project, filled with helicopters, sniper rifles, love, betrayal, and a whole lot of dead goats.

Platoons of professional goat-hunters scoured the Galapagos, searching high and low for the enemy. One by one, Galapagos goats caught a bullet to the heart or brain. Thousands of goat corpses were left to rot across the island, essential for making sure the total nutrients present in the ecosystem were not imbalanced via their removal (that would be a loss of biomass, a form of disturbance). By 1999, it was time to move into Phase Two.

As tends to be the problem with trying to eradicate *everything,* there will always be a few individuals that are especially good at not dying. The commanding officers of Project Isabella remarked that killing the final 5% of goats was just as difficult as killing the first 95%. It's easy to take down a whole crowd of goats milling about, but when their numbers get low, tracking down individual goats becomes a painstaking process. So the team employed a new tactic: a femme fatale honeypot, or as they described it, a *Judas goat.*

The process of creating a Judas goat is relatively simple. First, take a female goat and sterilize it. Then inject it with hormones that make it permanently in heat (estrus) and very, very attractive to males. Finally, drop your newly sexed-up Judas goat in the midst of an island's goat population. She'll be irresistible, drawing much of the local populace in like moths to a flame. Or like goats to anything remotely edible. Execute any suitors that appear, hide the bodies somewhere else on the island so as not to arouse suspicion, rinse, and repeat. It's diabolical. And if that whole plan sounds completely absurd, trust me, it was. It worked shockingly well.

By 2006, the Galapagos Islands targeted by Project Isabella were declared goat-free, save for a few Judas goats that remained, just in case. Other wild-roaming domesticated mammals, like donkeys and pigs, were dealt with in a similar manner as part of the project. This has not gone over entirely smoothly with the local populace. Apparently, some people on the Galapagos enjoy hunting the goats for their meat. Fishermen even attempt to re-introduce goats as an act of spite in the face of new regulations. If the balance cannot hold, the Men Who Shoot at Goats may soon be called back into action.

These are odd stories, but they bring up an important question. What makes us think we have the right to slaughter animals? There's a very pragmatic reason to do so: keeping a balanced ecosystem. Does that give us the right to go around gunning down, poisoning, or infecting populations that happen to live in what we've deemed the wrong place? We shouldn't use these tactics to deal with ecological problems caused by having too many *people* in one place, so why goats or rabbits?

"Hang on," I hear some of you saying. "Why is this any different from slaughtering cows and pigs and chickens by the millions?"

Factory farming is brutal as well. Killing animals for food poses different ethical questions, which have culture, tradition, and socioeconomics all mixed in. Culling invasive species belongs in its own ethical category, and I think the reasons for that have to do with human hypocrisy.

An uneasy familiarity enshrouds the subject of invasive species. Looking back on history both ancient and recent, we recognize that we ourselves are invasive. Since our origin in Africa, we've spread all over the world, driving many species extinct and critically endangering more in the process. Ironic that humans, one of the worst perpetrators of species displacement, are so set on keeping other plants and animals from displacing species. Not only that, we're *encouraging* that very behavior through our own actions. It's the species-wide equivalent of the way the United States handles global politics. "Do as we say, not as we do."

Biologists think of evolution in terms of species. With the exception of *Homo sapiens*, animals don't think in terms of their species as a whole. Selection works, for the most part, at the individual level. A single goat acts to benefit its own well-being, and possibly the well-being of its family or friends, but not the well-being of its species. How likely is it that a goat even understands the *concept* of a species? Or the concept of a *concept*? Even humans notoriously struggle with the biological concept of a species. We assign the species "goat" to an individual goat; the goat doesn't do so itself. Therefore, "goats" in the Galapagos must be blasted in the head with sniper rifles, because goats are harming other wildlife. It's easy for us to make that call. We're the ones with power. We're not being shot.

We may be doing what we think is best for biodiversity as a whole, but our track record on that front is not stellar. Humans have already dramatically altered the global environment, in some places almost beyond recognition. No amount of careful management on our part is going to restore the biosphere to what it once was; we can't turn back the ecological clock. Should humans try to "undo" the changes they caused, even when our incomplete understanding of ecology means they might be playing with fire?

We cull goats, toads, and rabbits en masse, but they didn't do anything "wrong." Right and wrong have nothing to do with it; each individual just tries to survive. Why do these animals have to suffer? Are their deaths a kind of reparation for the organisms they displaced, or is this ultimately about ensuring the continuation of the human species, by creating a more stable biosphere? If *that's* the case, then should we really be trying to control invaders?

Yes. Of course we should. If biodiversity is to endure, then we have to take action to control the invasive species we've unleashed, and that includes ourselves. But there are right and wrong ways of going about it, morally. These are a couple of the right ways to do it.

Killing an animal is different from stopping it from reproducing. The latter may still involve ethical issues, but it's certainly better than slaughter. Instead of killing the goats, we render them or their offspring sterile. The goats could still have sex, but it wouldn't take long for the population to age and die. With no offspring, the goat population gets exterminated just as completely, though over a longer period of time. The culling process on the Galapagos was already incredibly labor intensive. Sterilizing goats on the Galapagos would be more "humane" than going out and shooting them individually.

Another promising idea in ethical species control involves modifying a native predator or prey species. In this scheme, bringing one population's numbers down also brings another's up. This worked great in Yellowstone, where the white-tailed deer population got so out of control that many plants simply couldn't cope.[33] The reintroduction of wolves to the park had an immediate effect, re-balancing the deer population and

---

[33] This is probably a good time to mention that many places that Americans dubbed as "natural" areas, like Yellowstone and Yosemite, were actually carefully managed by the Native Americans who lived there until white people decided they knew better, threw the inhabitants out, and started screwing things up.

allowing native plants to grow again. We could also train native predators to consume invaders. Perhaps, instead of using bio-terrorism, we could train some entrepreneurial marsupials to go after Australian rabbits.

We have to accept the fact that there is no restoring an ancestral habitat to its "pre-human" state. Instead, it falls to us to be good caretakers of the human-altered world. Having gotten all of these species into this mess, it's up to us to come up with solutions that don't involve mass extinctions. It won't be easy, but with enough research and careful planning (far more than we've mustered to this point), humans could set altered ecosystems back into relative balance.

If we're smart, we'll learn from the past. Native Americans and other Indigenous peoples spent thousands of years practicing environmental stewardship, learning to manage the land. If we could recover and/or promote those that retain this knowledge, maybe we could incorporate indigenous management techniques to keep invasive species from getting too numerous.[34]

Stewardship gives us hope. With a lot of informed nudging, perhaps we could design habitats for all types of species — native and invasive — to live with one another, just as the house sparrows learned to live alongside us. The only constant for organisms, after all, is change. Rather than trying to go back to the "pre-human" biosphere, we can put our best

[34] For example, Native Americans often employed controlled burns, which could be used to quell invasive species that are not adapted to fires. One reason why forest fires are a huge problem on the West Coast is because colonizers stopped performing controlled burns, allowing forest debris to build up and fuel massive blazes.

foot forward, allowing as many different species as possible to live together in the brave new world we help create.

Like a symphony slowly building in volume and complexity, morality emerges, ever more complicated, in the brains of highly intelligent social animals. It's a biological synthesis of evolutionary common sense, empathy, and socialization. In this section, we went from the fringes of unconscious "functional" morality in drongos and meerkats and fungus, to ideas of reciprocity and fairness in monkeys and corvids and parrots, to empathy and compassion in rats, dogs, and chimpanzees. The same biological principles of natural and kin selection that led to the evolution of cannibalistic woodpeckers and spiders that offer themselves up as post-coital snacks *also* led to the evolution of a bonobo capable of looking into the eyes of a crying human and putting her arms around the person, as if to say, "It's okay, I'm here." Evolution has created elephants who grieve over lost loved ones and rejoice at reunions with friends and family after months apart. It has created dolphins that guide stranded humans back to shore. These behaviors are not at opposite ends of a spectrum — both work to further the propagation of genes in populations over time. Good and bad, sadness and joy, selfishness and altruism: they're all cut from the same biological cloth.

We looked for evidence of Darwin's proposed evolution of morality and found a trail leading all the way back through history to the beginnings of multicellular life on Earth. If Darwin could see the sheer volume of data we now have, charting the development of morally adjacent behavior in different species, along with our own, I bet he'd be overjoyed by how relevant his views on the evolution of morality remain. Humans don't have an "X-Factor" that leads to our heightened sense of right and wrong — it's just an extension of something present in many other species.

Yet, humans are still an enigma. Our brains are extraordinarily complicated. Our means of communication are scarily diverse. We exhibit a profound degree of empathy and mutual connection that pushes us toward "higher" moral behavior — what Darwin might call the evolution of the Golden Rule. We are inspired by the story of a person with no position, education, or power, someone who advocated the fundamental idea of unconditional love, forgiveness, redemption, turning the other cheek, and turning swords into plowshares. *Homo sapiens* is no "killer ape."

We have yet to fully understand our own uniqueness. Knowing where we came from is a big step toward knowing who — or what — we are, but it isn't enough to explain fully why humans act the way they do, especially in large numbers. Though evolution is less obvious to us when observed over thousands rather than millions of years, humans are still in a constant state of change. By synthesizing the fields of animal behavior and cultural anthropology, we can finally address what was Darwin's greatest challenge to explain: ourselves.

Part II: Leaving the Garden

16. The Making of Homo Sapiens

"We must, however, acknowledge, as it seems to me, that man with all his noble qualities ... still bears in his bodily frame the indelible stamp of his lowly origin."
— Charles Darwin, *The Descent of Man.*

What was supposed to be a chapter in the Descent of Man has turned into its own volume. Charles' newest project on human and animal emotions is progressing in a very different manner from his previous books: never has the subject matter been so close to home. To understand and explain human emotions, he must look not outward, but inward.

His friends and family members are study subjects now. Darwin diligently notes the emotional highs and lows of the family dog. He carefully observes the contortions of his son Francis' mouth while he plays the flute, contemplating the function of music for eliciting emotion in humans. He gazes thoughtfully at his own bearded face in the mirror. He looks tired. He is tired. But he has a book to write.

Letters begin to flood in from acquaintances who describe odd or intriguing expressions they'd witnessed in animals and people. For his section on emotional expression in infants, young mothers in Darwin's social circle report in with their own scientific observations– children pouting, laughing, crying, and the contexts in which they do so. Emotions are contagious, and so too, apparently, is the desire to study them. Many of these community-submitted observations make it into the final manuscript.

Interspersed throughout the book are professional close-up photographs of faces rendered into the whole spectrum of human expression. Joy, anger, fear, laughter, disgust, sadness, contemplation– they all pop from the pages on faces that are sometimes exaggerated, other times genuine and moving. The book venerates something very dear and fundamental to our humanity. Arising from these voluntary and involuntary muscular movements are stories, and they radiate from each and every face.

The book is published in 1872, entitled "On the Expression of the Emotions in Man and Animals." Emma is quite happy about the book's lack of controversy. In fact, the general public seems to be quite excited about understanding their own emotional repertoires. The book sells even better early on than the Origin of Species.

Darwin's Expression contains perhaps an even deeper concept. The beauty of the natural world is on full display in tropical rainforests and coral reefs. Yet that same natural wonder lies within each and every one of us – it is a beauty and complexity that humans carry wherever they go. Because Nature is not only something we observe. Nature is something we are.

///

Let's face it: humans are weird. They are bipedal primates lacking in the hair department, and are only good at four things: running, throwing, sweating, and inventing capitalism. We have already looked at plenty of strange animals, using science to understand how these biological oddities developed in the first place. Now, it's time to look in the mirror.

Primates, including all monkeys, apes, and hominids—diverged from other mammals right around when a dinosaur-erasing meteor hit some sixty-five million years ago. Six million years ago, hominids split off from the common ancestor we share with chimpanzees and bonobos. We're pretty sure that even this early on, hominid brains had already evolved to be relatively large, due to their subjectively advanced, cooperative social structure.[35] Everyone wants to know how they measure up in the social hierarchy. Early humans, like many other social species, kept careful track of their friends and enemies for this reason.

Four million years ago, the lush African jungle that early hominids called home was like a Garden of Eden, filled with greenery and tempting fruit trees. Intense climate change turned our "garden" into a sparse savannah, forcing us from the trees down to the ground, where we learned to walk. Modern humans are just one species descended from that walking primate.

Three million years ago, our plucky band of cooperative, tool-making, omnivorous African primates entered the Stone Age — stone tools became commonplace in everyday life. Two million years ago, hominids began perfecting one of the deadliest tools in their evolutionary arsenal: jogging. Using our shock-absorbent legs and state-of-the-art sweat glands, we went out in teams and trotted after prey until they collapsed from exhaustion.

Not long after, we gained control over fire and began cooking our food. Grilling meat, mushrooms, and tubers over the fire enabled us to get more calories from them, calories that

---

[35] Just as with the vampire bats, the size of the primate neocortex (which does a ton of the higher processing in the brain) is *directly* related to the size of a primate species' social group! Our social systems are why hominids probably got such large brains in the first place.

could be spent making better tools, and forming bigger brains. By around 500,000 years ago, Earth had at least three hominid species living simultaneously. *Homo heidelbergensis* were tall spear-throwing craftspeople, and the first to build elaborate shelters out of wood. *Homo neanderthalis* were stocky, art-making cave hunters of the icy north. These hominids lived along some kind of "beta" version of *Homo sapiens.* Anatomically modern *Homo sapiens* emerged some 300,000 years ago, but even before then, we probably shared our sister species' complex culture, artisanship, and maybe even spoken language. Sadly, those other hominid species that lived before or alongside us died out. *Homo sapiens* became the last surviving hominids on the planet.

We're the final children of a lonely branch on the tree of life. Why were we the only ones to make it out alive, reeling from our own near-extinction? There are biological and anthropological explanations that suggest it wasn't mere luck.

One idea is that humans required fewer calories than, say, *Homo neanderthalensis,* which was much larger. As a result we were much better suited to surviving the ice age, when the availability of food resources plummeted worldwide. Maybe this size difference also led to greater pressure for the young *Homo sapiens* to innovate. We couldn't stand our ground and stab a giant sloth the way a big beefy neanderthal could, so we had to come up with different tactics. Necessity was the mother of invention — better throwing spears, better gathering equipment, group-based hunting and gathering, and so on gave us the upper hand. Or... disease or infighting claimed the lives of our hominid cousins, leaving humans the last ones standing–an endangered species clinging to life.

Many of our species' most definitive behavioral traits were probably shaped by this time of extreme scarcity and environmental hardship. Humans developed an arsenal of morally adapted behaviors in order to survive the shifting climate. Two major characteristics made *Homo sapiens* stand out: our ability to cooperate and sacrifice for what we considered our "in-group," and our ability to rain focused, vitriolic wrath on whatever we considered our "out-group."

We've all, to varying degrees, experienced the weird duality of living with other humans. On one hand, there are people who — for no personal gain — were there for us during times of crisis, gave us resources, cared for us when we were sick, and so on. In the news and throughout history, it's not hard to find stories of people heroically putting themselves in harm's way or sacrificing themselves to save others. Humans will risk everything, throwing themselves onto a grenade, in front of a gunman, onto the subway tracks, or into a burning car, sometimes to save the life of a complete stranger. Heroes can come from anywhere.

Charles Darwin was intrigued by the ideal of human self-sacrifice:

> It is summed up in that short but imperious word *ought*, so full of high significance. It is the most noble of all the attributes of man, leading him without a moments hesitation to risk his life for that of a fellow-creature; or after due deliberation, impelled simply by the deep feeling of right or duty, to sacrifice it in some great cause.

Darwin's right. It's touching to see what people will do for one another, the lengths we will go for people we don't even know. We have our glorious moments.

On the other hand, human atrocities are *also* common — mass shootings, civilian war casualties, and so on. They're plagued by interpersonal violence, but often it's connected to a larger social or cultural violence. Ordinarily kind people can be made to do horrible acts under the right circumstances. That's no reason to condone or tolerate villainy. To come to terms with bad behavior toward others, though, it's helpful to understand some of the underlying biology.

Our evolutionary past substantially influences our day-to-day behavior. The same can be said for our ancestors' pasts. Obviously, the record of our ancestors' lives is incomplete, but it's fascinating how much we've been able to piece together. Here's what we know about our anatomically-modern ancestors:

1. ***Homo sapiens* developed an extraordinarily complex system of non-verbal communication predating the existence of language**.

Humans are expressive, empathetic beings and *need* to be around other people. Prolonged isolation can lower human lifespans and harm mental health. The drive to socialize is biologically tattooed all over our bodies, from the unique whites of our eyes, which make gaze-following easy, to our innate communication system of gestures and facial expressions.

Frans de Waal describes how our many subtle and non-subtle facial expressions, gestures and posturing tell entire stories. For example, covering our face with our hands is often a sign of shame and embarrassment — we don't want others to see expressions like blushing (a clear and honest evolved signal of embarrassment), which gives away just how awful we feel.

Chimpanzees do something similar, covering their mouths to keep others from detecting their "fear grin," a sign of submission. When we get the job or win the big game or get a text from our crush, we triumphantly raise our arms to make ourselves look bigger. What we're saying with that pose is "RAAAAH! Look at me! I'm dominant!" Athletes do this all the time.

On the flip side, when our prom-posal gets rejected or we learn that our goldfish died, we look "deflated." The next time you watch a game of basketball or football or doubles ping-pong, take a look at the differences in body posture between the winning and losing sides. The story of the game is written all over them. When we laugh or smile, our eyes narrow at the corners, suggesting that we are being genuine and feel comfortable since we're intentionally limiting our field of vision. The list goes on. Our body and facial expressions are sophisticated, and we're very good at reading people. Facial expressions are also instinctive, to a degree. This explains why people born blind can smile and frown. Blind humans use facial expressions and gestures the same way seeing humans do, and for the same reason: our ability to communicate with one another, as well as our desire to do so, is *innate*.

We are primed for communication and cooperation. Our bodies are equipped with the physical and behavioral tools to make it happen. We probably used facial expressions and body language to impart key information to our fellow humans in pre-linguistic times: "Quiet!" "I heard something." "I found the nest." "Don't eat that, it's poisonous!" "You need to back off." "How did the hunt go yesterday?" "Does this fig leaf make me look thin?" An eyebrow raise, a clenched jaw, a shrug of the shoulders – all of these basic facial expressions and gestures can sometimes say more than an entire verbal conversation.

2. Pre-agricultural *Homo sapiens* lived together in close-knit cooperative bands of up to 150 people.

Like many of our great ape cousins, humans evolved to live in close-knit groups composed of family members along with some more distant relatives and in-laws. Our ancestors benefitted from sociality. Darwin writes:

> The social instinct seems to be developed by the young remaining for a long time with their parents; and this extension may be attributed in part to habit, but chiefly to natural selection. With those animals which were benefited by living in close association, the individual which took the greatest pleasure in society would best escape various dangers, whilst those that cared least for their comrades, and lived solitary, would perish in greater numbers.

In other words, it's likely that some natural selection took place that rewarded our ancestors for keeping families together — the "familial group" probably became a consistent trait among hominids. In the face of howling blizzards, hungry predators and widespread famine, evolution rewarded hominids who banded together and made sacrifices for their kin.

Some of the strongest evidence that early humans lived in familial bands comes from the existence of menopause in human women. Menopause raises an odd question: why would the human body *intentionally* limit its ability to reproduce, when reproduction is what evolution is all about? Kin selection is the culprit. Giving birth is an exhausting and dangerous ordeal. If older women "turn off" their ability to reproduce, they simultaneously extend their lifespans *and* stop competing with their own offspring for access to mates. In return, grandparents are free to help care for their grandchildren, boosting the reproductive success of their offspring. Scientists dubbed this "The Grandmother Hypothesis."

Scientists have discovered that humans aren't the only animals that go through menopause — some whales do, too. Like humans, whales are long-lived, and their offspring require an extraordinary investment. Many whale pods are led by older "matriarchs," non-breeding females who lead their children and grandchildren to ancestral hunting grounds, memorized through years of experience. With grandma to help look after the kids, whale daughters can afford to have more babies.

It was probably the same in humans. Seniors are often looked to as both sources of wisdom and caregivers for young children. This resonates for me. I was lucky enough that my grandparents played a huge role in helping to raise me. Whenever I needed them, I could always count on my grandparents to pick me up from school, take care of me when I was sick, and give me advice.

Other great ape elders seem to take on this caregiving role as well. Retired former alpha male chimpanzees will take to playing with and watching over the youngsters. It makes sense that grandparents and occasional great grandparents serve as the anchors for human cooperative bands, helping to care for children while younger adults conduct more intensive labor. Before there were villages, it took a tribe to raise a child.

Cooperative human familial bands grew in size over the millennia. From 100,000 BCE to about 10,000 BCE, the average human group was probably somewhere around 100-200 people. That number is crucial for understanding human morality and behavior. Many modern hunter-gatherer groups, like the Yanomami people in the Amazon, usually peak at around 100-150 members. Psychologist Robin Dunbar also popularized this group number with his studies on Hutterite religious communities in the U.S., whose groups would start to split once their numbers exceeded 150. Dunbar believes that the human brain can't really keep track of social connections and relationships beyond about 150 people, thus we tend to split up once groups reach that size, and 150 has been dubbed "Dunbar's Number."

There might be other evolutionary reasons as well. Science writer Michael Shermer notes:

> It turns out that 150 is roughly the number of living descendants (children, grandchildren, great-grandchildren) a Paleolithic couple would produce in four generations at the birthrate of hunter-gatherer peoples. In other words, this is how many people they knew in their immediate and extended family, which is corroborated by archaeological evidence from the Near East showing that agricultural communities typically numbered about 150 people ... When groups get large they split into smaller groups. Why? The answer is moral discipline and behavior control.

So 150 people ends up being about the number of extended family members that a reproductively successful hunter-gatherer would expect to have and keep track of. Maybe we keep dividing ourselves into groups of that size because *that's* what our ancestral social, cooperative, and moral instincts evolved to deal with.

Humans have long lived in cooperative bands - many hands, working together. Photo by Mariano Cecowski. Accessed via Wikimedia Commons.

Understanding our social and cerebral limitations can help us organize ourselves more efficiently. Shermer continues:

> Studies on optimal group size (in terms of finding a balance between autonomy and control) by the military during the Second World War found that the average-size company in the British Army was 130 men, and in the U. S. Army it was 223 men. The 150 average also fits the size of most small businesses, departments in large

> corporations, and efficiently run factories. A Church of England study, conducted in an attempt to balance the financial resources provided by a large group with the social intimacy of a small group, concluded that the ideal size for congregations was 200 or less. The average number of people in any given person's address book also turns out to be about 150.

Humans work best when they nest groups within other groups. It's one of the primary ways that we make our super-densely populated lives livable. We form lifelong attachments to our social groups, making sacrifices and caring for one another. We use communication systems to cooperate within these groups, working as a team to survive and thrive. Yet, when it comes to interacting with groups that are *not* our own, the situation can be radically different.

As strong as in-group bonds are among humans, out-group resentment can be just as impactful. An out-group is "not us." It is the "other." It may be a rival family, sports team, or political organization. Dog person vs. cat person. East Coast vs. West Coast. People are constantly looking for ways to distinguish between groups. Often these distinctions are harmless enough– other times, however, they lead to dehumanization, where we fail to see other people as people at all.

Modern-day competitive sports, politics, and war are filled with dehumanization tactics like name-calling and symbolic association. These tactics exist to make conflict "easier", increasing the motivation of the in-group to confront the out-group.

Humans are psychologically averse to harming and especially killing members of their own species. It's easier to injure or kill someone if you truly hate them, especially if you don't even think of your enemy as human. Language is an effective dehumanization tool. We call people we don't like "cockroaches" or "maggots," annoying pests that we have no qualms about killing. Learn to associate a fellow human with a cockroach, and you've made it all the easier to drive a spear through them. Aldous Huxley, author of *Brave New World*, wrote: "The propagandist's purpose is to make one set of people forget that the other set is human. By robbing them of their personality, he puts them outside the pale of moral obligation." This is why countries try so hard to dehumanize their opponents in war.

Is our species evolutionarily primed for conflict and war? Did in-group/out-group discrimination evolve during a resource-scarce past where we had to fight over food and territory? Or is it just a neurological fluke? The scientific community is hotly debating this issue. It may be that conflict is very old, but large scale war as we understand it is new– a consequence of recent land-holding, farming lifestyle changes. Regardless, ingroup/outgroup discrimination is an ancient part of human history, and is essential for understanding the behavior of our species.

Nothing brings people together like mutual hatred of a common enemy. It is the great and unfortunate irony of our cooperative human existence: strong internal bonds, visceral outgroup revulsion. Sometimes, we can't agree on who goes in the in-group and who goes in the out-group. That kind of disagreement is at the root of many major social problems around the world.

3. *Homo sapiens* brains feature relatively new software on ancient hardware, and they love to take shortcuts.

While our ancestors' brains were large by animal standards for many millions of years, the human brain experienced its greatest growth over the last few hundred thousand. Brains make up two percent of our body mass, but twenty percent of our energy consumption, much more than for any other animal. Human brains are big, energetically demanding, and filled with biases about the world around us. In order to process information more quickly and efficiently in a complex and challenging environment, our brains became masters of association. Unfortunately, the quickest conclusion isn't always the best one. Snap judgments exacerbate the "us-versus-them" mentality at the root of prejudice.

One study found that flashing the face of someone of a different race, even for less than a second, activated the amygdala in subjects, which is the area of the brain associated with fear. Same-race faces did not result in the same amygdala activation. The greater the implicit race bias of the participant, the greater the fear response.[36]

Implicit Race Association Tests work by asking participants to match a face with a word. Some words are positive, others negative. Subjects are assessed based on how quickly they make the associations between faces, which alternate. In one Harvard study, participants are asked in one round of testing to associate white faces with positive words like "joy" or "triumph" and black faces with negative words like "sorrow" or "disaster." The associations are flipped in the other

[36] Which means "bias toward members of a particular race."

round, where black faces go with positive words and white faces go with negative words. The first race to have "positive" associations is randomized for each participant to alleviate bias. Generally, American Caucasians that take these tests tend to have a harder time associating black faces with positive words; it takes them longer to make the association, and they more readily make errors by attributing black faces with negative words.

These results do not suggest that we are born racist. Instead, a lifetime of conditioning from our upbringing, media exposure, communities, belief systems, and a myriad of other factors are at work. This conditioning takes advantage of the pre-existing structure of our brains to make snap decisions. We don't need to have a bad experience with something to be afraid of it — if enough people tell us it's bad, there's a good chance that we can be made to believe it *is* bad. We're not so different from crows seeing caveman masks.

Learning to make associations is a survival tactic for animals, because there are many mistakes that don't afford the luxury of learning from them. It's hard to remember that a mushroom will kill you if you're already dead. In the wild, learning to associate stimuli with danger or safety can be the difference between life and death. Such snap judgments lead to unintended consequences, sometimes violent ones.

Brains can be taught to associate animals or objects with danger, even when that association is illogical. There are countless phobias: spiders, narrow places, falling asleep, rain, cheese, et cetera.[37] People often know that the phobia they experience is irrational but are still unable to handle the

[37] Fear of cheese is called Turophobia.

emotional response that it provokes, and they panic. I don't have a phobia per se, but I absolutely despise house centipedes. I hate the way they rocket around on the walls, and their alarming number of long legs. I know I should like house centipedes: they are great pest controllers and harmless to humans, but I just can't go there. With a *thwack,* my sandal leaves my many-legged bedroom intruder dead as a doorknob. Our brains are constantly looking for threats and tend to err on the side of speed over rationality: better safe than sorry.

Human evolutionary history created a brain that is not hardwired, but incredibly susceptible to learning associations that are harmful to ourselves and others. We are often taught apathy toward one group of people, sympathy toward another, and fear and hatred toward everyone who isn't "us." Our cultural biases can dramatically change how we think and feel, making them a powerful tool for both collective action and collective manipulation. Are humans who are conditioned from a young age to think a certain way "at fault?" It's an uncomfortable question with no clear answer. Regardless, it's important to realize that our brains are vulnerable to bias and manipulation.

We'll get back to the brain soon. Suffice it to say, this organ is key when we talk about the link between biology and morality. Behavior is based both on genes and environment. Since our minds are so sensitive to the complex cultures we grow up in, the brain's ability to alter behavior is dialed up to eleven in *Homo sapiens* and other primates. To find more evidence, all you need is two monkeys, a cucumber, and some grapes.

17. Monkey See, Monkey Considers Moral Implications of Situation

"Origin of man now proved. Metaphysics must flourish. He who understands [the] baboon would do more towards metaphysics than Locke."

— Charles Darwin, personal notes

His friends are outraged on his behalf. After multiple years of being nominated but ultimately passed over for the Royal Society's most prestigious Copley Medal, Darwin has finally won! But there is a catch. Darwin receives the award "for his important Researches in Geology, Zoology, and Botanical Physiology." The Origin of Species, however, is not mentioned. In fact, one of the Royal Society Council members, Edward Sabine, states at the medal acceptance ceremony that they "have expressly omitted it from the grounds of our award." Darwin is not there to attend, but his supporters are. They know Darwin's landmark book is controversial, but an official endorsement of Natural Selection by the Society would be the real reward here, not the medal. They're not happy.

Ever Darwin's bulldog, Thomas Huxley protests the perceived slight at the reception, causing a stir. It's well-known that some of the Council members were harsh critics of Natural Selection. Back at home, Darwin receives a letter from another Council member, who tells him discreetly "Your friends– including myself did not fail to stand up for the 'Origin of Species' – as establishing a strong claim." Darwin had won the award with twelve out of twenty votes. It was close.

Charles' health is still declining– he is not able to work for very long without becoming extremely fatigued and prone to spells of confusion. One particularly bad spell leaves him unable to speak and with brief memory loss. He soon recovers, thanking God. He would rather die than lose his mind. These spells make it harder and harder for him to travel, indeed he rarely leaves home at all. Instead his home has been converted into an ever expanding laboratory– a research station of solitude.

Yet, Darwin is resolute in maintaining his correspondence with his many researcher friends. Though he is not there on the debate floors, his friends and supporters are. They ensure that his legacy is not lessened by some of his disgruntled colleagues in the Society. Darwin is not a boisterous man. He is not confrontational. But he has an ability to make friends who will stand up for him– an ability that will serve him well for the rest of his days.

///

When I have guests over and the conversation hits a lull, one of my favorite YouTube videos to show people is Frans de Waal's experiment on fairness with Capuchin monkeys.[38] In just a few minutes, this video explains a lot about our own behavior. It's the perfect way to steer the conversation toward something I'm interested in, without boring people to death with bird facts.[39]

The experiment goes like this: two capuchin monkeys in separate cages are assigned the task of giving the researcher a

[38] To watch this yourself, search "de Waal capuchins" on YouTube.

[39] I'm a PhD student who studies birds. I have many bird facts.

pebble. If they do, they get a slice of cucumber as a reward. Both monkeys are happy to do the task. They simply give up the pebble for the cucumber payment, and all is well. Then, the researchers make one critical change: rather than a cucumber slice, one of the capuchins receives a grape instead. What does the other monkey care? She's still getting a cucumber slice with basically no effort, so ... no harm no foul, right? Wrong. Upon seeing the other monkey being given a grape for the same task, the capuchin inspects her pebble. Maybe there's something wrong with it? She looks to be mulling it over. Could it be different from the one the guy next to me had? When the capuchin does the task again, and gets offered a cucumber again, and then sees her cagemate do the same for a grape — *again* — she flips out. She picks up the cucumber and tosses it at the researcher. She rattles the cage and slaps the floor with a look of sheer indignation. She is livid.

"This is basically the Wall Street protests, what you see right here." says de Waal. The crowd laughs nervously. I hope my party guests laugh.

These capuchins' behavior is far less extreme than that of the cannibalistic woodpeckers we met earlier, but both are based on a similar pretense: complex social systems require some degree of fairness and/or mutual benefit. The mad capuchin still gets a reward for the task, but the reward is now unequal, thus creating tension in the social group.

Interestingly, when this same experiment is conducted with chimpanzees, there's an added effect: the chimpanzee *getting* the grapes gets upset on their neighbor's behalf. The better-treated chimp recognizes the inequality of the situation, and she isn't happy about it — the stability of the social group is threatened. You don't want a jealous colleague clobbering you over a little grape inequity.

Above: these capuchin monkeys make sure to share. Photo by Frans de Waal. Accessed via Wikimedia Commons.

Systems of fair play exist all over. These systems are vital for group cohesion and were doubtlessly selected for over time in social animals. Species like crows, chimpanzees, and whales even pass down social and cultural practices that help ease tensions and cool tempers. Social animals mostly avoid brutality. Instead they cooperate, resolve conflicts without bloodshed, form lifelong friendships, and make personal sacrifices for the good of their kin and beyond. Human morality is no exception: nature selects for social cohesion.

Many human ideas about fairness are packaged in the form of religious beliefs: "an eye for an eye," "do unto others as you would have them do unto you," and so on. Darwin's critics and even some modern thinkers fear that losing a God-fearing moral center will cause society to devolve into sheer brutality, "every man for himself." These fears are misplaced. The Golden Rule isn't just in books — we see it play out in the behavior of all sorts of animals, plants, and even bacteria. Reciprocal altruism is *everywhere*. Religion isn't the *foundation* of morality, it's an *amplifier*. Let's see how it works.

18. A Moral Evolution

> "The highest possible stage in moral culture is when we recognise that we ought to control our thoughts."
>
> — Charles Darwin, *The Descent of Man*

In spite of many gloomy pictures painted by social media, humans are still profoundly moral animals. Moral ideas pervade our myths, are passed on from parent to child, and ring through our collective consciousness. Some moral impulses, like caring for children, are genetic, ingrained in our DNA through natural selection. Others, like giving money to charity, may be based on personal or historical experiences. Varied ideas of right and wrong emerged over millennia across all of our many cultures.

Moral diversity isn't surprising: humans now live all over the planet. The Earth harbors prodigious biological and geographic variation. To adapt to these places, our ancestors didn't just need to change the clothing they wore, the animals they hunted, or the crops they cultivated. Their moral senses had to adapt, too. Adapt, they did.

Consider the moral ideals of "individualism" (usually from the West) and "collectivism" (usually from the East). Individualism prioritizes the needs and goals of the individual, along with the rights to pursue these needs and goals. In theory, the pursuit of each person's goals benefits the group in the process. Conversely, collectivism emphasizes the needs of the group and what is best for the collective. Schools of philosophy and economics have been founded upon these moral ideas. Religions developed. Wars were waged — lots of wars. It's important to understand where concepts like "individualism" or "collectivism" originate.

Anthropology and sociobiology research has been telling. Take the Han Chinese, one of the oldest extant human civilizations, considered to have a "collectivist" culture. Peace and harmony are widely promoted, particularly ideas of sameness. Of course, China is a massive place and harbors immense cultural variation, but certain moral ideas work to unify the country. The idea that everyone must work collectively for the benefit of the group might be traceable to the grain that Chinese people have farmed together for thousands of years.

In a published paper in *Science*, a team of researchers led by Dr. Thomas Talhelm interviewed over 1,000 college-age Han Chinese people. These subjects were assessed psychologically on collectivist versus individualist ideals. How do you measure collectivist vs. individualist thinking? Apparently, one way is to give subjects a set of words, such as "bus," "train," and "tracks." Try it for yourself. Which *two* words do you immediately associate more closely? If you associated "bus" with "train," that suggests analytic, or "abstract," association. Bus and train are in the same category: vehicles. This kind of association is more frequent in individualistic cultures. Associating "train" with "tracks" is a holistic or "relational" association — "train" and "tracks" are in the same category because trains move along tracks.

Try it again with "rabbit," "carrot," and "dog." Individualists tend to associate "rabbit" with "dog" because they are both animals, whereas collectivists tend to associate "rabbit" with "carrot" because rabbits eat carrots. Participants were given many questions along these lines, and the researchers found that Han Chinese people from communities that traditionally grew rice overwhelmingly answered in a collectivist way. At least, that was true for the *southern* Han Chinese subjects. For *northerners*, there were slightly different results: they trended toward individualism. The reason? Talhelm believes it might be traceable to the different types of crops that northerners and southerners traditionally grew. While the Southerners farmed rice, the Northerners farmed grain, and the different processes involved apparently led to cultural and behavioral differences.

As Sapolsky explains in *Behave*, different farming styles can drastically change the social structure of a community. Rice paddies require massive, collaborative irrigation efforts. Irrigated water flows into everyone's paddy, so everyone has to contribute to the cultivation of the entire field. This is a moral philosophy made manifest: in order to grow your *own* rice, you have to help grow *everyone's* rice. It stands to reason that communities that relied on rice had to adapt their moral values to reflect the requirements of their farming practices. Communities that tried to grow rice *without* collectivist moral values would probably have been out-produced.

In contrast, wheat can be grown independently. It requires less water, grows in open fields, and needs fewer people for cultivation. Wheat farming encourages greater independence, individual focus, and a bit less reliance on the community. Historical crops seem oddly effective in predicting moral values.

The researchers in this study didn't just test word association. In economic games, those from rice-farming communities were *less* likely to "punish" (take money from) those who were previously dishonest or unfair in games, thereby avoiding conflict. Subjects also drew "sociograms," which are diagrams of the subjects' relationships. The diagrams are represented by interconnecting circles, including the research subject.

People from rice-farming areas drew circles for their own names that were smaller than the circles of their friends. Those from wheat-farming areas drew their own circles 1.5 millimeters bigger than those of their peers. According to *Scientific American*, Europeans draw their personal sociogram circles 3.5 millimeters bigger, and Americans draw themselves *6 millimeters* bigger.[40] A pattern emerges. The more individualistic a culture, the bigger the self-value of its individuals.

What's more, most of the test subjects weren't personally associated with rice or wheat farming at all. Industrialized agriculture has substantially lowered the number of people needed to cultivate crops, and yet the cultural ideas that come with rice farming persist in populations that traditionally farmed rice. Ecological and agricultural backgrounds can directly influence human behavior toward others.

Weather is also a major influence on behavior. Take siestas—those short midday naps taken after lunch. Siestas may seem like a non-sequitur, but consider when it might make sense to take a break from work in the middle of the day. It's not laziness. Civilizations in hot and/or humid climates, like the Mediterranean or the tropics, are precisely those that take siestas or midday breaks. Laboring away under the sun when you live *that* close to the equator is a great way to get

---

[40] For my fellow Americans out there, 6 millimeters is 0.23622 inches.

heatstroke. That's why the Anglo-American 9-to-5 work schedule doesn't fit in every climate where it's implemented.[41]

Cultures, customs, and traditions can all be impacted by a number of ecological factors. We are coming to a tricky topic, namely cultural evolution. To review: biological evolution states that, over time, species descend with modifications resulting from natural selection, mutation, and so on. *Cultural* evolution suggests something similar: over time, human cultures are selected for, modified, and even adapted to the circumstances in which they find themselves. Many moral beliefs were selected for. The fields of biology and cultural anthropology intersect here, but the two don't always mesh well.

Darwin understood the potential of cultural evolution to modify behavior, providing multiple accounts, in *The Descent of Man*, of what we might call "cultural traits" and their potential adaptive values. Of course, Darwin didn't know about DNA, so he didn't distinguish between biological and cultural traits. Still, it's clear throughout *Descent* that Darwin had a kind of cultural evolution in mind, at least for humans.

Unfortunately, cultural evolution is mostly devoid of physical evidence. How can one show that cultural evolution took place? We have "fossil records" of sorts, in the form of stone fragments, tools, and ancient scrolls, which help us piece together stories, but it's up to us to decipher them. In biology, though, the evidence of evolution is everywhere in the form of DNA. We can analyze it, sequence it, and even modify it to suit our needs. But what's the DNA of culture? How do we pass

[41] The American industrialist Henry Ford tried to recreate a midwestern town in the middle of Brazil called "Fordlandia."

cultural ideas and the ingrained knowledge of our ancestors down from one generation to the next? Simple: stories.

Stories have power. They allow us to imagine problems, confronting them in our minds before they ever take shape. They define our social environments, giving us moral information about the actions of those around us. Stories are how we effectively encode not only morals, but information altogether. Like DNA, they get passed onto future generations, and can even mutate over countless retellings. It is because of their versatility, their capacity to completely reshape how we think, that stories are the cornerstone upon which our species is built.

In his book *Cultural Evolution*, the researcher and author Alex Mesoudi describes a study in which participants were told to remember a set of facts. They were later tested on their ability to recall the facts accurately over time. Different groups were presented with the information either in a straightforward manner or in the form of a story. Consistently, groups that were presented with story-based information recalled that information more accurately, and for longer. Psychologist Jerome Breuner found that facts told as part of a story were *far* more likely to be remembered. That might be why religious figures like Buddha or Jesus are famous for speaking in parables.

Our minds are hardwired to absorb information from a narrative. It's often said that the best works of fiction contain elements of truth. There's a reason why people so often ask or tell children what the *moral* is to a story. As our cultures develop, we continually come up with more vivid, more complex, more varied ways of expressing cultural ideas. What are cultural ideas? They are the ever-changing stories we tell about ourselves.

Beginning in the Stone Age, humans likely used verbal storytelling, music and cave painting for both entertainment and passing on information. The earliest cave paintings date back to before 30,000 BCE. Over time, storytelling media expanded to include plays, clothing, pottery, books, merchandise, even silverware. These media communicate history and identity, contributing to that larger amalgam of ideas that comprises a "culture."

Written language was world-changing for many civilizations. Before writing, it was much harder to keep a record of what had been done, who owed what to whom, and what the gods willed. Writing allowed cultural information to be stored and retrieved at a moment's notice. A literate person could learn the *Epic of Gilgamesh* by dusting off a stone tablet rather than tracking down the village bard who mostly remembered the story. Those stories could become longer and more detailed, making epic sagas more accessible than ever before. Writers traveled across time and space, producing works that spanned generations and crossed continents.

As the historian Yuval Harari describes it, writing, along with money, "broke the data processing limitations of the human brain." Stories became consolidated, and universalized, as they were written down. Government officials could keep elaborate records of sales and purchases, population censuses, and mandates from the higher-ups. In fact, the extensive record-keeping of the Sumerians is why we know so much about them today. Writing made possible the organization of people into larger and larger bureaucracies, linking chains of command via tablets and scrolls passed over vast distances. Knowledge could be secured in written form, dutifully copied down by a scribe or librarian. Organization and unity were made possible on never-before-seen scales.

Darwin writes in *The Descent of Man*, that "As man advances in civilization, and small tribes are united into larger communities, the simplest reason would tell each individual that he ought to extend his social instincts and sympathies to all the members of the same nation, though personally unknown to him." We now know it's not always so simple. Plenty of forces can drive groups of people apart, and only a select few truly bring people together.

For our species, unity is power. Historically speaking, we know that the ability of groups to band together has made all the difference in the world. When humans unite and put their heads together, they can alter ecosystems, change lives for generations, and literally move mountains. And there seems to be no unifying force more influential than religion.

19. Are You There, Darwin? It's Me, God

"The impossibility of conceiving that this grand and wondrous universe, with our conscious selves, arose through chance, seems to me the chief argument for the existence of God."

— Charles Darwin, *Letter to N.D. Doedes*

"...The conviction of the existence of an all-seeing Deity has had a potent influence on the advance of morality."

— Charles Darwin, *The Descent of Man*

It always seems to come back to Him.

Plenty of interviewers and colleagues have asked Darwin about God. Now, he must ask himself. Summer rain patters against the windowsill. He's been writing an autobiography of sorts– something of himself that will be left behind for his family. Now it is time to write what will surely be the most controversial part. He's entitled the next chapter "On Religious Belief."

Some years before, Darwin and Asa Grey exchanged letters about how God might impact their scientific findings. The question was, to what extent were the various shapes and patterns of nature actively sculpted by a divine hand? Darwin was a believer when he departed on the Beagle and when he wrote the Origin of Species, but his views evolved with age.

"I cannot think that the world, as we see it, is the result of chance;" he once told Grey "and yet I cannot look at each separate thing as the result of design."

It is the paradox of the world. The beauty of nature and the horror of nature lie side by side. Beautiful rainforests and horrific slavery. Loving families and dying children.

"There seems to me too much misery in the world." He continued later. "I cannot persuade myself that a beneficent & omnipotent God would have designedly created the Ichneumonidæ with the express intention of their feeding within the living bodies of caterpillars, or that a cat should play with mice...[42] *On the other hand I cannot anyhow be contented to view this wonderful universe & especially the nature of man, & to conclude that everything is the result of brute force. I am inclined to look at everything as resulting from designed laws, with the details, whether good or bad, left to the working out of what we may call chance. Not that this notion at all satisfies me. I feel most deeply that the whole subject is too profound for the human intellect. A dog might as well speculate on the mind of Newton. — Let each man hope & believe what he can."*

This was his tried-and-true Deist approach: believing in a God but not one that actively intervenes in the world. Of course the true nature of God was beyond him. Of course it was. But... then, he'd added this:

"One word more on "designed laws" and "undesigned results." I see a bird which I want for food, take my gun & kill it. I do this designedly. An innocent & good man stands under a tree & is killed by flash of lightning. Do you believe... that God designedly killed this man? Many or most persons do believe this; I can't & don't."

---

[42] The *Ichneumonidae* are our familiar parasitoid wasps.

Perhaps this final anecdote hints at the true source of Darwin's frustration: not God, but the way religious institutions claimed to know God. According to many of them, God designedly killed three of Darwin's children. The Church commands considerable authority. He has been berated by many bishops and pastors. They've called evolution dangerous, immoral, and cruel. They've ridiculed and misrepresented him. Meanwhile, he has to sit there and take it when he's told that a baby's death from disease is all part of the plan. It's unbelievable to him. A loving God is no executioner of the innocent.

By 1876, his frustrations finally pour out in this autobiographical chapter. Writing it must feel cathartic. There are many words he's wanted to use for years, but couldn't. No longer.

Charles makes it clear that he is no atheist. He writes about his reluctance to give up his Christian belief– striving to find evidence that could continually convince himself. Eventually, Darwin realized that there were many interpretations of Scripture that he found horrifying. Darwin cannot reconcile how loving God is with how wrathfully He behaves– with interpretations of the Old Testament "attributing to God the feelings of a revengeful tyrant."

Hell is perhaps the most gruesome concept of all. "...The text seems to show that the men who do not believe, and this would include my Father, Brother, and almost all my best friends, will be everlastingly punished. And this is a damnable doctrine."

Not only does God see fit to take the lives of his loved ones on Earth, but so too must he punish them for an eternity in the afterlife? It cannot, Darwin thinks, be true. This doesn't mean that God does not exist. On the contrary. "...I feel compelled to look to a First Cause having an intelligent mind in some degree analogous to that of man;" he writes. "And I deserve to be called a Theist."

Later on, when Emma Darwin reviews the manuscript for post-mortem publication, she and others will elect to omit many of the more volatile parts of this chapter– including Charles' comments on hell and his 'revengeful tyrant' line. "Nothing can be said too severe upon the doctrine of everlasting punishment for disbelief–" she hand-writes in the margins "-but very few now [would]. Call that 'Christianity,' (tho the words are there.)" The full of Charles' writing in this chapter won't be made publicly available until 1958.

Husband and wife may disagree about Christianity, but they do agree about the importance of church. In the rural village of Down, the church is the cornerstone of community involvement and outreach. Charles serves on multiple church committees, overseeing charities for the poor and working with clergymen to manage finances for the local parish and fund services. In fact, one of Charles' closest friends in Down is John Brodie Innes, a conservative pastor with views radically-opposed to his own. They never see eye-to-eye on Christianity or evolution, but care for one another all the same. They work together on multiple charitable causes over the decades. Emma and their children attend services and volunteer at Sunday school. They visit people in need, bringing money and food. Darwin understands that the concept of a church is about more than just scripture and worship. Because no matter how he evolves– whether he is an Anglican, Deist, or Agnostic, he still believes in treating others the way he'd like to be treated. He still believes in the

Golden Rule. He still– in some way or another– believes in God.

///

"Religion" is a tricky term to define. Darwin's contemporary, Matthew Arnold, defined religion, in *Literature and Dogma*, as "morality touched by emotion." This is profound, but vague. *The Oxford English Dictionary* calls religion "the belief in and worship of a superhuman controlling power, especially a personal God or gods." This definition isn't great, either. Religion is a little more complex than just belief in a God or gods.[43] Despite my middle school teachers' insisting that Wikipedia is unreliable, I actually prefer *its* definition, which reads "a social-cultural system of designated behaviors and practices, morals, worldviews, texts, sanctified places, prophecies, ethics, or organizations, that relates humanity to supernatural, transcendental, or spiritual elements."

A complex definition for a complex term. After thousands and thousands of years of spirituality and worship, the origins of religion are obviously a bit murky. We can't really be sure *where* religion first developed, but observing the behavior of animals might give us an idea of *how* it got going. Like many good ideas, religion probably started when hominids began using visions, dreams, and drugs to enter altered levels of consciousness.

Humans are not the first animals to figure out that ingesting certain chemicals can result in a great time. The

[43] *Dungeons and Dragons - 5th Edition* defines "Religion" as a knowledge type skill linked to Intelligence. It can be used to gather knowledge about religious doctrine or to recognize a religion-related clue.

animal kingdom is filled with species that regularly use recreational drugs. For example, when they're not flying Santa's sled around the world, reindeer regularly eat magic mushrooms to get them sky-high. A reindeer on 'shrooms behaves weirdly, running around erratically and making uncharacteristic braying noises. When the feeling wears off, they will work hard to forage up more of the good stuff.

Teenage dolphins take turns passing around a pufferfish to get their kicks. Pufferfish are deathly poisonous, but in small enough doses their poison appears to produce a narcotic effect in dolphins. Likewise, some of our primate cousins, like lemurs, love to take careful bites of toxic millipedes to produce a similar feeling of euphoria. Humans aren't even the only animals who enjoy alcohol. Birds and monkeys get sloshed off fermented fruit. A group of cedar waxwings in Minnesota once got so wasted that locals started calling the cops on "rowdy drunk birds," as an NPR reporter put it, who kept flying into windows, falling over on the street, and being a general public nuisance. Whether the birds meant to get drunk or not, it was still pretty funny.

Humans have probably used recreational drugs and alcohol as long as we've been able to get our hands on them, and they play an important role in many world religions. Alcohol is everywhere in connection with religion.[44] Ancient Mesoamerican cultures like the Aztecs frequently used psychedelics in religious ceremonies. Nearly 4,000 year-old Vedic scriptures from India mention some kind of psychedelic

[44] Doing Passover "by the book" requires you to drink many cups of wine. Catholics drink Jesus' *blood* in the form of wine ... and so on and so forth. You could fill a lot of books with the ways religions have used alcohol.

drink known as "soma." A line from the *Rigveda* translates roughly:

> We have drunk the soma we have become immortal;
> We have gone to the light; we have found the gods..
> What can hostility do to us now, and what the malice of a mortal,
> O immortal one?

In the book *Inside the Neolithic Mind,* the archeologists David Lewis and David Williams Pierce argue that psychedelics probably played an integral role in early religious experiences. But it wasn't just drugs. Throughout history, humans have found ways of accessing different states of mind. Some spiritual experiences are 'hypnagogic', which uses the semi-lucid state right between being asleep and being awake to induce hallucinogen-like visions. Other experiences rely on rhythm, meditation, and even exhaustion. Learning to induce these altered states was important for members of spiritual groups, and may have even helped raise one's social status.

This holds true for many religious groups today. Lewis and Williams comment that "[Christian] monks through their chanting, fasting and incessant prayers were able to reach mental states considered closer to God than any that lay people could attain. Those few who could go beyond euphoria and dissociation to experience vivid visions were more likely to achieve positions of influence in the church and therefore in affairs of state as well."

In prehistory, it is possible that the myriad geometric patterns, sounds, and images seen while hallucinating may have actually served as useful inspiration for designs, inventions, art, and storytelling. This makes complete sense. In an altered state, the brain can absorb all manner of novel stimuli, opening up new possibilities for creativity. Even people who use psychedelics today describe cosmic, spiritual experiences that are life-changing. Here is a description of an angel from Ezekiel 1:15-21:

> As I looked at the living creatures, I saw a wheel on the ground beside each creature with its four faces. This was the appearance and structure of the wheels: They sparkled like topaz, and all four looked alike. Each appeared to be made like a wheel intersecting a wheel. As they moved, they would go in any one of the four directions the creatures faced; the wheels did not change direction as the creatures went. Their rims were high and awesome, and all four rims were full of eyes all around.
>
> When the living creatures moved, the wheels beside them moved; and when the living creatures rose from the ground, the wheels also rose. Wherever the spirit would go, they would go, and the wheels would rise along with them, because the spirit of the living creatures was in the wheels.

This passage– with its flying wheel made of eyes– certainly suggests that the narrator is experiencing an altered state of consciousness. The geometric patterns, otherworldly design, and feeling of reverence all feed together to create new, powerful imagery. This description of an angel sounds as if it were *experienced*, rather than imagined.

Modern religions are of course filled with cultural practices and moral guidelines that have nothing to do with altering consciousness, which are used for coming-of-age ceremonies, weddings, or just everyday life. It's possible that over time, the drug- or meditation-induced revelations of spiritual leaders merged with the practices and moral customs of cultural leaders, resulting in the hybrid moral-spiritual phenomenon that we know today as "religion."

Religion, in its many forms, has been at the center of countless major events in human history. Moral ideals associated with religion, such as pacifism, "God-given" human rights, the sanctity of marriage, the divine authority of emperors, and the lingering obsession over conquering Jerusalem have started uprisings, founded nations, destroyed nations, and changed the destiny of humankind. Given what we know about Neanderthal behavior and their possible ceremonial objects, and given how common psychedelic substance use is in animals, it's possible that religion predates modern humanity itself.

Early humans lived in small groups consisting mostly of close relatives. Part of religion's origin was likely in finding a way to cope collectively with the loss of relatives and honor the dead. Every culture on Earth has some form of ceremonial ritual for the dead, and these are usually connected in some way to that culture's predominant religion. The loss of relatives, especially children, has been a regular occurrence for all of human existence. It is morally important to every civilization on Earth that the dead be treated with respect. Religions, especially ones that promote the concept of an afterlife, offer some survivors a way to move on collectively.

Religious beliefs, rites, and ceremonies connect with other natural phenomena as well. These events include the changing of the seasons, migration of animals, and the ripening of crops. As touched on earlier, most weddings have religious undertones, as do coming-of-age ceremonies. Even the earliest forms of religion likely incorporated these important events into systems of traditions and moral beliefs.

Above: Religious practices play an important role in day-to-day life for billions of people. Photo by Sanjay Been. Accessed via Wikimedia Commons.

Pre-agricultural religions developed among small groups in near isolation. Organized religion emerged later when people began growing crops and living in larger communities. The Agricultural Revolution predated writing by nearly 7,000 years, and sophisticated belief systems began to develop during this period. However, it wasn't until the invention of writing that the true potential of religion to shape society was revealed.

Written documents allowed larger and larger groups of people to all embrace a single moral belief system. At the core of modern religions are texts: the Buddhist Tipitaka, the Christian Bible, the Taoist *Tao Te Ching*, and so on. These texts serve as the foundation for religious doctrines and beliefs, which are in turn expressed through objects and symbols — dress, food, architecture, and music, to name a few. Organized religion allowed large communities over multiple generations to unite under a common set of moral and spiritual ideals. Moral unification is a powerful tool — remember, we need ways to keep ourselves organized beyond groups of 150 people. The more coordinated and unified a group of people, the more profoundly they can impact the world around them, and the stronger they become collectively.

Religions inspire collective motivation to many ends, some peaceful, others violent. The world's most widespread religions — Christianity, Buddhism, and Islam — all have one idea in common: they all promise peace and everlasting happiness as the final reward for a life well lived. Of course, the definition of a "life well lived" varies depending on the religion *and* the denomination. Buddhists seek freedom from attachment and to follow the teachings of the Buddha as the path to enlightenment. Most Christians believe that their faith in Jesus Christ as their lord and savior will absolve them of original sin and permit them an eternity of joy in heaven. Muslims believe that those who do good deeds on Earth and are faithful to Allah will be rewarded for their deeds in paradise. Though these motivators differ a little, they're all appealing rewards.

If even those motivators fail, there's also the promise of an eternity in Hell. Christianity and Islam both did a thorough job of terrifying me about damnation. Hell is described as a land of unspeakable horror, filled with searing heat, agonizing cold, rivers of blood, and endless punishment. It's like Cleveland, but worse.[45] Human brains are even more focused on avoiding pain than seeking pleasure, and that's why the *threat* of Hell may be a stronger motivation than the promise of heaven. If there is a God, I suppose He's well aware of this.

The promise of reward or punishment in the afterlife can motivate people to extraordinary action, even superseding their own instincts of self-preservation. Who will be less likely to run from conflict, the warrior who thinks that death is the end of existence, or the one who believes that if he dies a glorious death in battle, he will be rewarded with an eternity in paradise? Viking berserkers believed they would be welcomed into Valhalla for their glorious deaths. Christian Crusaders believed their effort to retake the Holy Land would surely land them in heaven. The fiercest warriors we know of believed that they fought for an objective, divinely given moral ideal. The strength of their beliefs went on to alter the course of history through war and conquest. Even Genghis Khan, of all people, was seriously religious. When you are convinced that you're doing right by the gods, there's less room for doubt.

In the same way, people who think their bosses are jerks for giving them long hours will probably not work as hard as those who think the fruit of their labor is evidence of God's blessings and that laziness is an immoral trait of sinners. The moral holiness of productivity is a staple of Protestant faith. Many athletes and artists contribute their success to a higher power pushing them onward. Religion motivates people to undertake and accomplish things they would not otherwise.

---

[45] Both places feature flaming rivers as part of the scenery.

Charles Darwin also gave a lot of thought to the influence of religion on human morals and behaviors. In *The Descent of Man*, he writes:

> The motive to give aid is likewise much modified in man: it no longer consists solely of a blind instinctive impulse ... With the more civilised races, the conviction of the existence of an all-seeing Deity has had a potent influence on the advance of morality.

Darwin notes here that religion can deliberately serve to create positive moral and behavioral qualities in humans, qualities that are grounded under the watchful eye of God. If religious belief increased the human "motive to give aid," then it was a highly useful tool for societies indeed. To this day, many researchers view religious beliefs as biologically adaptive.

Biologist Donald Broom writes:

> If morality has evolved and morality is the core of religions then the basis for religions has evolved. All human societies have a propensity for religion because religion provides a valuable structure for the moral code which is valuable in all of these societies. The religious framework makes it easier for the average person, or perhaps more importantly the likely transgressors of moral codes, to understand what should and should not be done.

Over a century later, and we are still echoing the words of Darwin.

Religions reinforce cultural values, solidifying them with the backing of supernatural forces. Cultural norms and values differ from group to group, and religions vary in content. The moral needs of societies from times long past have made their mark on the practices of modern religions. It makes sense that, in the conquest-driven empires of antiquity, groups with religions valuing hard work and skill in battle outproduced and outcompeted religions lacking those values.[46]

Is "competition" occurring today between religions? Are Christianity, Islam, and Buddhism "winning" over different belief systems since they have the most followers? At a Bible study session in college, someone told me that the popularity of Christianity was evidence of its status as the "true" religion, stirring the hearts of a plurality of people. I don't view religions with more followers as superior.

While biomass has some use as a measure of species' success, it's bogus to rely on it in this instance. There are many historical reasons why certain religions have more followers. Factors like the population sizes of countries and forced conversions come into play. It's hard to argue that certain religions are "better" because the people that crossed the seas five-hundred years ago forced those they encountered to convert under threat of death. Here's another reason why numbers aren't everything: some religions are split up into so many smaller denominations that they should almost be counted as two or more. Given that Protestants and Catholics have slaughtered each other for centuries, perhaps "Christianity" is not extremely useful as an umbrella term.

Religions aren't necessarily competing; it's more accurate to say that religions developed in different moral and cultural

---

[46] That's with a level playing field though-- even if the enemy is afraid of war and kind of lazy, a zealous kingdom of ten-thousand shouldn't take on an empire of ten million based on faith alone.

environments to serve different needs (though there are obviously many human needs that are universal). Instead of asking which religions are "best," we should be asking why there are so many different, often radically contradictory religious doctrines in the first place. Looking at religion as a numbers game misses its most fascinating aspect: adaptability. In spite of strict doctrines, religious beliefs are in fact incredibly malleable when we think in terms of millennia instead of lifetimes. Adaptability has allowed religions to shape themselves in accordance with shifting ecological and social environments. Commonplace religions are filled with wildly inventive solutions to the challenges that come with civilization. Like the unusual adaptations in animals discussed in earlier chapters, religious beliefs and practices have a method to what occasionally appears to be madness.

Many religious customs seem bizarre out of context, but make good sense in context. In Judaism and Islam, eating pork is forbidden. For ancient civilizations plagued by plagues, that prohibition was totally logical. Pigs are frequently the source of zoonotic disease even today (think of the swine flu). The Old Testament God likewise compels Jewish men to be circumcised. Modern medical research has shown that circumcision lowers the risk of urinary tract infections, certain sexually transmitted diseases, and penile cancer, all on top of making the genitalia easier to wash. Today, the medical benefits of circumcision are mild, but "back in the day," when access to hygiene was much more limited, circumcision likely saved many lives.

Islam generally forbids alcohol. That's particularly interesting, since alcohol plays such a large role in the ceremonies of many other religions. It also happens that many of Islam's 1.8 billion followers live in areas with little rainfall. The Mecca, Islam's holiest city, receives almost no rainfall all year long. Alcohol is a diuretic, which takes water from the blood. The "hangover" that people feel after drinking results from dehydration. If water is *extremely* scarce in the desert, it seems logical to avoid dehydration, and therefore alcohol. Maybe people following this practice were more likely to survive in extreme desert climates than those who didn't. Perhaps this is a cultural, moral adaptation. Perhaps it is sheer coincidence.

Intermittent fasting is popular in health circles. A number of studies show that fasting has positive impacts on brain health, metabolism, even overall lifespan. Many religions incorporate fasting periods into their belief systems, such as Lent, Ramadan, or Yom Kippur.[47] Back before agriculture, we probably "fasted" naturally due to the nature of hunting and gathering — regular meals were less common. Our metabolisms are adapted for a "feast/famine" lifestyle. Maybe agriculture-based religions incorporated fasting as a way to help our bodies adapt to agriculture. It would make perfect sense. To sum it up, no religion or belief system is objectively "better," they simply have different advantages in different contexts, yet all share a common denominator of promoting social cohesion and stability.

Religion and "cultural adaptation" also intersect. Even the *belief* in a higher power and an afterlife could be culturally or biologically adaptive. Psychology studies show that people who devoutly believe in a benevolent, loving God tend to cope better with stress, exhibit fewer symptoms of depression and

[47] Check out Trepanowski and Bloomer, "The impact of religious fasting on human health" (2010).

anxiety, and respond better to mental health treatments. It's not hard to imagine why.

Being human has its perks, but among the biggest negatives — and positives — is knowing in precise detail what has happened, what is happening, and what may soon come to pass. Humans are distinctly aware of their impending deaths. Knowing you're going to die one day, and that there is nothing you can do about it, isn't necessarily a bad thing, but it can present some mental challenges. As a kid, I'd always "known" about my own mortality, but I don't think I became acutely aware of it until I was about fourteen years old. I remember the day it happened — I couldn't sleep the entire night, thinking about my impending demise. If I had been able to convince myself, at the time, that my soul would live on after death and be granted everlasting life, I think I'd have slept a lot more soundly that night.

The suffering of others bothers me, sometimes even more than my own. I feel more at peace thinking that those who have passed away — friends, family members, even people I don't know — have moved on to a better place where they can be happy together. It's nice to think that, in spite of how horrible a place the planet can be, there's a loving being watching over us, maybe with an ultimate plan. The *act* of believing in a loving God, afterlife, et cetera, seems to make life better for a lot of people.

Plus, religion gives meaning to life, not just in the way its adherents are meant to live, but in terms of what moral ideals they work *toward*. Some Protestant groups believe that the path to salvation lies in doing good works here on Earth, and that motivates them every day to contribute something good to the world. Buddhists believe the path to salvation lies in achieving enlightenment and freeing oneself from all attachments, and some spend a lifetime striving for complete mental peace. As long as they have hope and something to work toward, people have reason to persevere.

We've talked about the biological basis of our most basic emotional and "moral" responses to the world around us. We've seen how different evolutionary pressures led to different levels of cooperation, communication, and competition in other species, including our close relatives, the great apes. Could culture and its close cousin religion allow humans both to follow their biological moral impulses, and to modify, supersede, or even transcend them?

20. A Will To Change

"Hence if man goes on selecting, and thus augmenting, any peculiarity, he will almost certainly modify unintentionally other parts of the structure ... "
— Charles Darwin, *The Origin of Species*

Fame is beginning to take a toll on his private life. Down House is continually invaded by wayward visitors. They've made their pilgrimage to speak with the prophet himself– and the reclusive prophet and his family are forced to adapt.

Some of Darwin's visitors are expressly invited friends and celebrities, but many are complete strangers. The prolonged socialization is exhausting for him, but he sees taking visitors as his duty. Darwin stays as long as is socially-acceptable, then makes an artful escape, citing his health, work, naptime, or literally any other excuse. His family, with Emma at the helm, covers for him and entertains guests in his absence. They manage Darwin's endless stream of visitors like a business unto itself.

Young naturalists worship the ground he walks on. They come away from their brief meetings changed, starstruck by meeting the legend himself. One admirer is so overcome with emotion that he cries on the way home. The prophet is rapidly gaining disciples.

Attitudes toward evolution are softening as his fame grows. He is interviewed for books about great naturalists of the age. Even with all the affection and attention, Darwin wisely never beats his chest. He is quick to credit others and slow to credit himself. He works to maintain a reputation for gentle, quiet humility for the rest of his life– which probably only adds to Darwin's stardom. The biographer Francis Galton asks if he has any special talents.

> *"None," he writes in reply. "Except for business... Very methodical in all my habits."*
>
> *He must be the least-talented great naturalist who ever lived.*

///

Though most humans have completely reshaped the way they would have lived 10,000 years ago, our biological design is fundamentally identical to when we were still chasing animals across the savannah and foraging for mushrooms and tubers. Our paleolithic hunter-gatherer ancestors likely organized their systems of beliefs around many of the same principles that present-day non-farming, non-domesticating societies value.[48]

In societies that rely on hunting and gathering, spiritual beliefs tend to lean much closer to nature-centered Animism. The "gods," if present at all, are many. Alternatively, a spirit — sometimes called mana or orenda or simply the Great Spirit — inhabits all things. Many stories in these societies revolve around animals as characters, and humans are just another character in those stories. Animist stories often contain useful

[48] Here I run the risk of making the same mistake Darwin did. There is an immediate major flaw with this line of thinking, and it should be made abundantly clear: all societies on Earth have been developing for an identical amount of time. I bring these societies up because they bear a greater similarity to how we think early humans lived before agriculture, not because they are "primitive" or "backward." No group of people should be regarded as such.

lessons about the natural world. For example, in many Native American myths, Coyote is a trickster who will go after your stuff. Stories from Great Plains Native Americans about Buffalo describe his many "gifts" in the form of food, clothing, and shelter. Medicinal herbs are ascribed powers in their own respective origin stories. The stars have their own stories to assist in navigating and keeping track of the seasons.

Animist myths culminate in a spiritual view that places humans in intimate relationships with the rest of nature. Morally, animals and plants are to be treated with respect, because without this respect, one could easily find oneself facing down a panicked or hostile animal, or accidentally eating a poisonous plant or mushroom. Knowing the right way to interact with nature can save your life.

The San people of the Kalahari Desert still practice persistence hunting, as featured in the David Attenborough documentary *Human Mammal, Human Hunter*, in which the San run their prey down to exhaustion. Watching San hunters run makes me feel like a weakling. Presumably, this is the oldest form of hunting around and explains *Homo sapiens*' ability to run absurdly long distances.[49] The San people don't rely on their stamina alone.

At one point in Attenborough's documentary, while chasing after a kudu (a kind of antelope), one of the San hunters loses track of the animal when it runs into cover. With no trail to go on, he begins to imagine himself from the perspective of the animal, even holding up a hand to mimic the

---

[49] Humans are even known to outrun horses if the race is long enough. This is because horses aren't bipedal and waste energy running on four legs. "Why the long face, horse? Did you just get outrun by a primate or something?"

kudu's head movement and gait. He visualizes himself as the very kudu he is hunting, using his empathy and imagination to deduce where the kudu went. He soon finds it, kills it, and honors the animal with spiritual ceremonies. This Animist ideal — that of placing oneself in the position of the animal and identifying closely with it — is not only spiritually, but also physically important for the success of the group. The study of animal behavior may be the world's oldest profession.

Hunter-gatherer groups also tend to swing strongly egalitarian — in general, humans in groups numbering fewer than 150 tend to be. Anthropologists in the twentieth century found that, almost universally, hunter-gatherer groups tend to operate in close bands of around 20-50 or so. These groups often maintain friendly, sometimes familial relationships with nearby bands, and internally, they have strong beliefs regarding cooperation and "consensual decision-making," in which each member has a voice.

In hunter-gatherer cultures across Africa, Asia, and South America, individual equality is important. People distribute food resources more-or-less equally, regardless of who obtained that food. Materials are shared among everyone. Many major decisions are made collectively. These societies demonstrate measures which prevent more skilled and experienced individuals from using their prowess to gain advantage in the social hierarchy. In an interview, a !Kung hunter[50] put it this way:

> Say that a man has been hunting. He must not come home and announce like a braggart, "I have killed a big one in the bush!" He must first sit down in silence

[50] The "!" part of the !Kung name is a clicking vocalization.

> until someone else comes up to his fire and asks, "What did you see today?" He replies quietly, "Ah, I'm no good for hunting. I saw nothing at all ... maybe just a tiny one." Then I smile to myself because I know he has killed something big.

Another !Kung man, a healer, explains:

> When a young man kills much meat, he comes to think of himself as a chief or big man, and he thinks of the rest of us as his servants or inferiors. We can't accept this. We refuse one who boasts, for someday his pride will make him kill somebody. So we always speak of his meat as worthless. In this way we cool his heart and make him gentle.

The structures of these societies focus on promoting harmony and avoiding envy. That is not to say that present-day hunter-gatherer societies are free from conflict (many frequently have internal struggles or wars with out-groups), only that they possess a consistent set of ideals. Many anthropologists believe our hunter-gatherer ancestors held similar ideals to those of hunter-gatherers today.

The development of agriculture some 10,000 years ago was a catalyst for massive, widespread cultural and environmental change. Whereas hunter-gatherers move around, farmers stay put. With fixed property came homes and expensive tools, which required time and effort to make. With homes and tools, people were able to accumulate even more resources. Gone were the days of sharing everyone's resources among the tribe — our groups began to balloon way beyond the 150-person sweet spot. Living in permanent settlements meant that land became *owned*, rather than moved around on. Familial inheritance of land and resources became a priority.

Above: a group of San people building a structure. Photo credit: Andy Maano. Accessed via Wikimedia Commons.

Labor in agricultural societies is structured very differently from hunting and gathering. Small bands of hunters and gatherers with differential levels of success are replaced by repetitive labor, day in and day out. Farmers are at the absolute mercy of the weather: no rain, no food.

Agriculture brought with it the rise of powerful sky gods — Zeus, Indra, Asman, Amun-Ra, Odin, Anu, and Quetzalcoatl, just to name a few.[51] For crops to grow, these gods had to be appeased with sacrifices and rituals. Gods forbid that you make them angry. Spiritual agendas shifted from the land to the heavens. Human-god diplomacy became paramount. This shift appears in the relative stories and mythologies of hunter-gatherer and agricultural societies — when farming starts, the gods become more personified, and humans are created in the image of the gods themselves.

This godlike image of humans, or *some* humans, is important for agricultural groups.[52] The egalitarianism of hunter-gatherer societies gives way to staunch authoritarianism —emperors, kings, and god-kings — in virtually every ancient agrarian civilization we know of. Ancient Mesopotamian cities were built around temples. Gilgamesh, the hero of the oldest written myth ever found (dated some 4,000 years ago or more) is the king of Uruk: 50% man, 50% god, and 100% jerk to his subjects. Divine mandate is a very useful way to legitimize the rule of a monarch or emperor. The Ancient Egyptian Pharaoh went from being god-like, or an aspect of the gods, to a straight-up deity in the eyes of his subjects. His word was the word of the gods themselves, and as such, it had to be obeyed without question.

---

[51] These are examples of sky gods largely from civilizations later on, but versions of them no doubt existed in early agricultural societies.

[52] In ancient Egypt, for example, many of the gods were portrayed in an in-between stage, with the body of a human but the head of a bird or dog or crocodile. Others just looked like people.

Not all land is created equal — certain sites are better for farming than others and generate more fertile crops. Fertile land encouraged the transition to a much stronger social hierarchy. If a family or clan has claimed that land and has enough power to hold it, they have leverage over those who wish to work that land. It's not the *farmer's* land they are working, it's the chieftain's ... or the king's.

Also, as discussed earlier, permanent sites are hard to guard. They're often rich in resources and invite interlopers and conflict from other groups. In order to keep intruders out, one's people must be trained in hand-to-hand combat and archery. Due to the surplus goods and people generated from farming, agricultural societies could afford to redirect bodies away from food production. Adam Smith wrote about specialization in *The Wealth of Nations* in 1776, and he was right: with farming, people can afford to specialize in and dedicate themselves to other tasks, such as priesthood, craftsmanship, and of course, war. Hunter-gatherers were already engaged in some specialization, but agricultural societies took it to a whole new level.

Distinct classes and castes begin to emerge sometime after 10,000 BCE. Around the world, from the southern tip of Argentina to the icy shores of Svalbard, the archeological record shows social stratification in farming communities. Early hierarchies generally had a peasant or laboring class, a warrior class above them, and then a religious, ruling class. The ruling class, by way of the warrior class, usually gets exclusive control of force in feudal agricultural societies. Warriors get the swords and armor, while the peasants get pitchforks. In addition to fighting wars over territory and resources, the warrior class is there to provide "protection." Protection often involves keeping the peasants in check and quelling revolts against the ruling class, in exchange for better treatment.

Apart from the change in management, agriculture brought another huge change to the human environment: the sheer number of humans and animals living alongside one another. To review, social animals benefit from the protection of groups, but they also suffer from increased risk of disease and greater in-group competition for mates and resources. As group sizes ballooned, cultural adaptation was key. Circling back to our earlier discussion of religion, belief systems that emphasized purification and cleanliness were probably healthier when living closely with more humans and livestock.

Jealousy and conflict arise easily. When everything isn't shared, theft, murder, vendettas, and other blood-soaked activities start to make their way into the culture. When civilizations are young, they struggle with the challenge of so many people and their respective conflicts. Another adaptation is needed: a code of laws to provide specific guidance as to how individual behavior should be judged. Stealing someone's goat gets your hand cut off. Poking out someone's eye gets your own eye poked out. Say what you want about Hammurabi or the Old Testament God, at least the rules are clear.

Once invented, agriculture spread rapidly across the planet. Similar patterns of social stratification and hierarchy exist even in civilizations that developed in complete isolation from one another. For example, the Mesoamericans and the Sumerians never met, but both developed similar farming techniques. Cultural and technological changes have consequently accelerated in our species. Like a boulder rolling down a mountain, this pace of change gains momentum with each passing second. The implementation of large-scale farming catalyzed a giant chain reaction – creating technological innovations, widespread conflict, industrialization, and a moral reckoning for *Homo sapiens.* Our species blinked and found themselves transported to a world so technologically advanced that we are now threatened with extinction due to our ecosystem-destroying levels of success. The evolution of *Homo sapiens* took 3.6 billion years. How might we survive the next two-hundred?

Part III: The Moral Anthropocene

21. Driving, Genes, and Gene Drives

"If the misery of the poor be caused not by the laws of nature, but by our institutions, great is our sin."
— Charles Darwin, *Voyage of the Beagle*.

Science at Down House is infectious. Many members of the general public yearn to share their own biological observations with Darwin. This is also true for his children. One of the aging naturalist's greatest joys is working with his offspring on experiments. He and his now-adult son Francis have set up a laboratory of sorts at home. Together, they excitedly measure the movement growth of seedlings toward light sources and the speed at which plants respond to touch. At the time, it isn't widely believed that plants can 'interpret' and react to outside stimuli. Darwin and son are showing that plants are much closer to animals in this respect than previously thought.

Their instruments are cobbled-together, made with whatever spare parts lay on hand or refashioned from existing pieces. They peer at samples using an ancient microscope and use his chemical balance from primary school. It is this characteristic jerry-rigged setup that defines the older biologist's technique– he is from the era of "gentleman science."

Darwin's methodology isn't long for this world. A new scientific order is emerging at the end of the 1800s: science that is institutionalized, formal, and industrial. Quaint country houses and makeshift equipment are being replaced by large laboratories where experiments are tightly controlled and researchers closely supervised.

It is a massive cultural change for science.

The person who helped bring about this new formal era of biological science is foundationally part of the old school. He refuses to buy fancy new gadgets. Later, Francis will write "I have always felt it to be a curious fact, that he who has altered the face of Biological Science, and is in this respect the chief of the moderns, should have written and worked in so essentially a non-modern spirit and manner." Francis has been to the modern laboratories and seen the new techniques, but he admires his father's approach just as much.

Visitors to Down House– especially botanists– marvel at the simplicity of the Darwins' setup. Darwin and son are making important discoveries in the field of plant biology despite their rudimentary instruments. It is like seeing a wooden sailboat hold its own in an age of steam-powered ironclads. As the two continue to make discoveries, they show that ingenuity and clever, simple experimental design are more important than all the new lab equipment in the world.

Darwin's old ways still have some fight left in them yet.

///

We began our journey all those pages ago in search of characteristics unique to the human species. I would like to add another feature to the mix, though it's more of a philosophical idea than a biological fact. Humans have a unique ability to transform their wildest imaginings into reality. We can make something real just by believing in it hard enough. For example: money. Money is technically imaginary. The little pieces of paper and metal we carry around and exchange for goods and services aren't actually very useful on their own. Dollar bills only have value because a critical mass of people *believes* they have value. If we all collectively stopped believing a dollar had value, the global economy would collapse. That is our species' power of faith and imagination. Some people say it makes the world go 'round.

Our cultures are made of stories. Stories come from both our experiences and imaginations– they often mix the real and the fantastical. Abilities that were once confined to the realm of our dreams and fantasies – flight, electricity, thinking machines, space travel – have since manifested into reality. It's a wild time to be alive. Fiction has transformed into nonfiction. Our technology seems closer and closer to magic. It's all happened so fast that we are woefully unprepared for the consequences.

Again, humans are anatomically the same bipedal primates we were some 100,000 years ago. The difference is we now possess weapons and technologies that surpass pagan gods in their destructive capacity. We have the technological means to tear down rainforests, punch holes in our ozone layer, and fundamentally shape life as we know it. For 3.6 billion years, our ancestors were at the mercy of natural selection and sheer luck. Today, we sculpt life on the genetic level. Humans are now the most influential ecological factor on the planet. Our species will continue to shape fiction into reality, growing more powerful in the process. It's our job to figure out how to use that power responsibly, that is to say, without devastating our populations and our planet.

Science can't help us much here. William Jennings Bryan, a lawyer, politician, and evangelical Christian who is featured prominently in the infamous Scopes Monkey Trial, argued the same: science can do a lot for us, but it can't necessarily tell us what is morally good. Humans will have to reconcile powerful new technologies with their ever-varied, ever-shifting standards of right and wrong. Easier said than done.

More technological issues are crossing the moral horizon. Over the past two decades, our understanding of and ability to manipulate DNA has grown tremendously. In 2013, CRISPR was first used to edit the genome of human cells.[53] During the so-called "Green Revolution" at the end of the twentieth century, agricultural output spiked, due largely to the genetic modification of crops. By messing with the genes of corn, soy, and apples, we were able both to improve food outputs and to make plants that we like resistant to herbicides that kill only the plants we don't like. In the United States, an estimated 85% of all harvested corn is genetically modified. Soon that number will be nearly 100%, given the yield advantage of genetically modified corn. Growers without those modifications won't compete in the long run, save for selling to a small group of anti-GMO purists.[54] We still grow plants out of the ground, but modern day agriculture is otherwise nearly unrecognizable from the days in which we planted and plowed by hand and horse.

Yet another promising avenue in genetic experimentation is the "gene drive." This technique can genetically modify entire populations of wild animals by introducing modified organisms with "super-heritable" genes, which have more than the normal 50% chance of being inherited. These genes get into the population over time as organisms reproduce, until eventually nearly everyone has them. Gene drives could help control

---

[53] CRISPR stands for "clustered regularly interspaced short palindromic repeats."
[54] GMO stands for "Genetically Modified Organism."

mosquito populations or other invasive species, limiting their ability to reproduce or harm specific other species. In our continuing effort to adapt to our changing environment, gene drives could even be used to give plants and animals resistance to pesticides, or to take it away. The possibilities are many, and often worrying, because just about *every* big project we try has unintended consequences. What if people begin to intentionally introduce detrimental genes into populations of crops or livestock as an act of bioterrorism?

Humanity's ability to manipulate genes will continually improve as time goes on. One day, people may be able to bend the fabric of life however they wish. The ticking time bomb of technology is just as real for our biotechnology as it is for artificial intelligence —once again, it is not a question of whether, but when these technologies will be fully applied.

With great power comes great responsibility. When it comes to taking responsibility for great power, the human species is a bit like a fifteen-year-old who's just been put behind the wheel of a Ferrari. Emerging genome editing technologies will be used in ways that make us confront new, unorthodox moral quandaries. In fact, we're confronting them now.

In 2018, the world came face-to-face with its future when a Chinese scientist named He Jiankui made the first genome-edited human babies. He wasn't creating superhumans, mind you: his goal was to make the babies HIV-proof. HIV (Human Immunodeficiency Virus) works by targeting the immune system; it hijacks white blood cells and uses them to make more viruses. Attacking the immune system is a brutal tactic, effectively wiping out the body's primary means of defense against the virus in the first place. HIV itself doesn't kill, it's the resulting AIDS (autoimmune deficiency syndrome) that allows all manner of pathogens to infect the body. With no immune system to fight them off, those pathogens can eventually kill.

Jiankui's idea was this: modify the CCR5 gene, which controls the structure on the outside of white blood cells, to change its shape. Doing this might prevent the virus from attaching to white blood cells, since the "password" for getting inside them would be changed, effectively rendering the virus harmless. In-vitro fertilization is needed to make those modifications, with test subjects carrying their modified babies to term.

In secret, Jiankui modified the embryos of couples who were HIV-positive. The virus is easily passed from parent to child, and so these children were at risk of contracting HIV. Shaky ethics to say the least! To start, Jiankui didn't receive clearance to undertake genome editing on human subjects. He and his team evidently used a forged ethics approval document from the Shenzhen HarMoniCare Women and Children's Hospital to conduct their work. The "informed consent" forms that his subjects signed were later found to be incomplete and not up to minimum standards of disclosure.

After the *MIT Technology Review* broke the story on Jiankui's work, the scientist revealed the fruits of his labor to the public, eventually presenting his research at the University of Hong Kong in 2018 when two babies, twins, were born. The story was picked up by more and more news outlets. *The Telegraph* referred to him as "China's Dr. Frankenstein."

The very next day, Chinese authorities shut down Jiankui's lab and detained him. Jiankui was later found guilty of forging documents and breaking ethical rules. He was sentenced to three years in prison and fined three million yuan, or close to half a million US dollars. The babies seem to be doing just fine.

This incident ignited a firestorm of ethical debates over human genome editing. The World Health Organization called for an end to such research and began actively tracking its occurrence. Jiankui's work was also denounced by bioethicists far and wide, even condemned by the Chinese Academy of Medical Scientists. Drafts of legislation aimed at curbing further human genome editing sprung up in countries around the world, China included, and the World Health Organization has since put out a moratorium on editing embryos.

But the genie is out of the bottle. How will Jiankui be viewed in fifty, or even one hundred years, when human genome editing is commonplace? Will he be seen as a mad scientist, a visionary willing to take huge risks to do the "right thing," or somewhere in between? How will we feel, in the future, about the first genetically modified fetuses of 2018, once we're modifying fetuses as casually and often as we modify corn today? Time only knows.

In the meantime, it's important that humans think critically about the impacts that genetic modification will have on society and, of course, on our senses of right and wrong, our morality. So much of the human experience, from our height to our struggles with disease, is affected by our genetics. Genes conferring athleticism or disability could completely change someone's life. Some genetic traits, like the ones affecting skin pigmentation or hair shape, are used as identifiers for cultures and communities. What happens when genetic predisposition is no longer based on ancestry, but on "choice?"

Homo sapiens should act quickly and decisively to control the course of genetic research and development. The effective "breeding" of humans harkens back to the days of chattel slavery or Nazis. What happens if genetically programmed elites figure out how to cheat the system? What if genome editing causes unforeseen genetic defects or abnormalities in the long run? The genetic lottery might be unfair, but rigging the results may prove disastrous. People have talked about the dangers of genetic engineering for a while now, and some warn about alarmist reactions. Philosopher Neil Levy writes:

> Genetic engineering arouses a particular horror in many authors ... It should be clear by now that this fear is misplaced. If genetic engineering of human characteristics proves to be possible and practicable ... it will not endow us with powers that we do not *already* possess ... If we fear that it will give the rich access to enhancements that are not available to the poor then we need to recall that such enhancements are already widely available to, and widely used by, the wealthy- in the superior schooling, nutrition, medical care, and other advantages that they can buy.

It's true that genetic engineering will probably work with existing human genes, but the issue becomes *who* gets access to this technology. Getting a rare useful or harmful genetic mutation isn't fair, but at least it's mostly outside of human control today — when designer babies emerge, it'll be a different beast. Even Levy's own argument seems to suggest that genetic modification will just lead to further differences between the haves and have-nots, exacerbating already widespread inequality.

That said, Levy also has a point. It's easy to focus on the drawbacks or dangers of a new technology, but we also need to think about the potentially life-changing benefits of genome editing. The work of Dr. Jiankui was conducted in a poor and unethical fashion, but the fundamental idea that we should genetically modify babies to be resistant to HIV certainly doesn't *seem* unethical.

Hereditary diseases are a good place to start in weighing the ethics of genome editing. Many people are born with hereditary medical conditions that, even with treatment, can make life extremely difficult. Consider sickle-cell anemia, which is caused by having two abnormal copies of the hemoglobin-making gene, leading to sickle-shaped red blood cells. Symptoms include anemia, pain attacks, infections, and even stroke. The condition affects millions of people, especially in Sub-Saharan Africa, where 80% of cases occur.

One explanation for the persistence of sickle cell disease is that one *normal* copy of the hemoglobin gene, paired with an abnormal copy, generates a partly altered blood cell believed to confer some natural resistance to the malaria parasite. Thus, doing some basic Mendelian genetics, if two people who are "carriers" of sickle cell have a child, that child has a 25% chance of having normal hemoglobin (no resistance, but not life-threatening), a 50% chance of altered hemoglobin (malaria resistance), and a 25% chance of sickle cell. In evolutionary terms, it may have been that the 50% chance of resisting malaria and 25% chance of sickle cell was worth it in the long run. This is why the trait has persisted in Sub-Saharan Africa where malaria, to this day, kills over 1.6 million people *per year*, the vast majority of them children.

Given the means and lack of side effects, would societies have a moral obligation to genetically modify people with sickle-cell disease, giving them a single normal copy of the hemoglobin gene? At the very least, parents could be given the option to choose gene modifications for their children. Perhaps they would even have the option to *give* children a copy of the abnormal gene that on its own confers malaria resistance or some non-heritable equivalent. Over time, gene editing of this kind could potentially save millions. If individual human lives are ultimately valuable, as most people believe them to be, those numbers are pretty convincing.

Other genetic diseases are every bit as destructive. Conditions like Huntingtons, Alzheimers, Muscular Dystrophy, and Cystic Fibrosis impact the quality and length of life for millions. In fact, nearly one in five people may carry genes associated with a major genetic disease. With the right grasp of genetics, we could wipe those diseases off the face of the Earth, eradicating them as we did with smallpox, freeing the world from their horrible influence. Once societies have the *means* to eliminate these diseases, they will likely feel an obligation to use them.

Researchers are already delving into this. In the case of Cystic Fibrosis, scientists have developed special proteins that can find a mutation in the CFTR gene (which is responsible for Cystic Fibrosis), correcting the mutation in living cells. This hasn't been tried in people yet, but it seems to work in the lab. With Alzheimer's, a number of genes and corresponding proteins seem to heighten or lower the risk of disease. The first clinical trials in live patients, using modified viruses to deliver corrective genes, are just getting underway.

Even if we are effectively "playing God" by altering the human genome, it might be acceptable to do so for painful, life-threatening, and quality-of-life-affecting genetic disorders. But even focusing only on diseases leads us into uncertain territory. If we make it our mission to stamp out genetic disorders, who determines what constitutes a "disorder?"

Down Syndrome is a genetic condition caused by an extra chromosome. The condition can cause delayed development and intellectual disability, ranging from mild to more severe. It's common — some five and a half million people have Down Syndrome worldwide. Is it ethical to eliminate the extra chromosome during or before pregnancy? Do we consider neurodivergence and a lower life expectancy sufficient grounds to alter or eliminate an individual's genetics? What about the millions of people who currently live with the condition? This moral dilemma is nothing new — many pregnant people carrying fetuses with Down Syndrome choose to terminate their pregnancies. Since Down Syndrome became detectable by early screening, the termination rate lies somewhere near 70%. This is not an easy moral choice, but it is one that people continually make.

Autism is a neurodevelopmental disorder with genetic components. Depending on the individual, symptoms range from difficulty with social interaction to an inability to live independently. Though severe cases can be debilitating, many of the world's greatest artists, thinkers, writers, and scientists are autistic. According to Dr. Darold Treffert, one in ten people with autism show extraordinary abilities in music, mathematics, and more, along with a working memory that exceeds that of most neurotypical people. If we can screen and modify the genetics of zygotes, should individuals have a right to modify development in order to "avoid" neurodivergence? Should pregnant people have the right to choose not to carry a potentially autistic fetus to term, or to modify its genes instead? The moral value of the autonomy of pregnant people could conflict with values of neurodivergence.

This book does not offer definitive solutions to these challenges. A rough and rocky moral road lies ahead, no matter how technologies like artificial intelligence and genetic modification move forward. As science provides us greater power to shape the world around us, we face moral challenges with greater long-term consequences. There is no getting around these issues. The future is upon us, and figuring out what *should* be done with the mythology-scale abilities we are acquiring won't be easy.

22. Adaptation, Across the Nation

"How paramount the future is to the present when one is surrounded by children."
— Charles Darwin, *Letter to W.D. Fox*

In 1871, Charles Darwin wrote that "the social instincts ... with the aid of intellectual powers and effects of habit, naturally lead to the golden rule, 'As ye would that men should do to you, do ye to them likewise.'" Darwin could not have known just how world-changing that statement would be, nor could he have known how deep and varied human moral beliefs truly are. Though beliefs concerning right and wrong vary from culture to culture, we all ultimately have a lot in common. I've tried to point out over the course of this book that social norms, values, and ethics translate into real-world selected-for behaviors in humans and other species. "Moral relativism" is the idea that moral judgments are right and wrong relative to a particular culture, time period, or situation, and that no one moral belief is inherently better or worse. That's not completely true in the case of humans. Reviewing human history and identifying values, human societies demonstrate strong universal themes about right and wrong; it's the minutia we tend to fight over.

We should not assume that human beings are the only species capable of acting morally. For example, the beauty and importance of an individual's newborn children seems universal in humans. Whales, elephants, and great apes would probably agree. After all, we are all K-strategist mammals with very long gestation periods. Each baby constitutes a monumental effort on our part, and it is vital that we protect it.

Imagine an advanced civilization on a planet orbiting Proxima Centauri — the closest star to our sun. Let's say this civilized species evolved from R-strategists that acted a bit like crabs or insects. If this species laid thousands of eggs at a time and only a few of them survived after hatching, their moral system probably wouldn't be super adamant about the sacredness of babies. It seems likely that if these crab aliens developed abortion technology, it wouldn't be nearly as contentious an issue on their hypothetical planet as it is on ours.[55] The crab alien lawmakers would probably say something like, "Who cares? These eggs are a dime a dozen — let's eat!" And that would be that.

Darwin takes it a step further in *The Descent of Man*, suggesting that if our ancestry were different, we would even consider the commitment of what we primates call "atrocities" to be perfectly normal. He writes:

> If ... men were reared under precisely the same conditions as hive-bees, there can hardly be a doubt that our unmarried females would, like the worker-bees, think it a sacred duty to kill their brothers, and mothers would strive to kill their fertile daughters; and no one would think of interfering.

Asking which moral belief system is best is a bit like asking what gene is best. It depends on the context. The counterargument is that certain genes are absolutely fundamental to the success of pretty much all living beings. Our biology is why, though human groups can differ, so much

---

[55] I never, EVER thought I would be stringing "crab", "alien," and "abortion" together in a sentence. Yet here we are.

of our morality is universal: some moral ideas are just plain pragmatic. For instance, a society that believed that newborns should feed themselves wouldn't last for very long.

If you want to talk about universal morals for humans, it pays to be deliberately vague. Still, all humans have a shared biological and social ancestry, so some consensus about broad moral concepts exists. For example, humans are hardwired to cooperate, even if that cooperation is mixed in with competition. We evolved to work together, and it makes sense that we feel some moral imperative to do so.

Humans all value the lives of their children. Not other people's children, that varies — but our own? Yes. When infanticide does happen in human cultures, it is almost always out of desperation. It's like Ghandi said. "Humanity is an ocean; if a few drops of the ocean are dirty, the ocean does not become dirty." Legend has it that the Spartans left their weak and sick babies to die on a cliff. This isn't true. Even Spartans loved their babies.

Humans also all believe in something larger than themselves and ascribe some aspect of "right"-ness to this belief. This could be a religion (in fact, all known societies *ever* encountered have some form of spiritual belief and/or practice), philosophy, ideology, currency, or any thought that extends beyond the material world. In Western culture, these types of concepts — often in combination — manifest as "God." God isn't necessarily "a figure in the sky who talks to humans," but a guiding force behind many people's everyday moral actions. Even the most materialistic people ascribe *some* value to immaterial ideas like artistic expression or love, however small.[56] That, in and of itself, is a beautiful feat of evolution. Or, at least, another happy little accident.

There are more universals out there. Michael Shermer lists over 200 moral and religious universals for humans, including marriage, sanctions against offenders, taboos against incest, admiration of gift giving and generosity, and so on. We agree on a lot. It is much more productive to try to find common ground than to focus on how we differ. Just as important as the morals we currently *have* are the morals we will *need* for the future. Of course, the two may be one in the same. A penguin that adapts its wings to function as fins works with what it has available. So does a whale with its air-breathing lungs. Now our moral and cultural ideas, built up over generations of human struggle and sacrifice, must adapt, too.

In 1949, the French philosopher Simone de Beauvoir published *The Second Sex*, a landmark text of second-wave feminism. In it, she argues that the biological limitations imposed by childbirth made women economically dependent on men, and this dependency deepened over time as the effectiveness of certain technologies and weaponry corresponded with the user's physical strength. The stronger you were, the better you could produce or fight, so the theory went.

Of course, the rise of industrialization changed the technological landscape of the world. It also has increasingly made physical strength less relevant in terms of economic productivity. De Beauvoir argues that mechanization has made obsolete the old social structures which oppressed women and

---

[56] Even cold, diabolical Bond villains have an appreciation for fine art and classical music, as well as giving long monologues to the hero so they have just enough time to escape.

relegated them to menial tasks. In order to achieve social equality for women, she writes, women will need to secure economic equality, made possible by technology.

All kinds of organisms retain "obsolete" features. Evolutionary biologists sometimes refer to "vestigial structures" when describing organisms' anatomy. A vestigial structure is a body part or organ that does not seem to have any current function. The design of the vestigial structure, though, suggests that it was useful in the past.

For instance, snakes have little bumps of bone where their legs used to be. Evidently, they got rid of legs altogether, preferring a more grounded approach to movement. Blind fish that evolved to live in pitch-dark caves have eyeballs despite their uselessness, a remnant of a time when they could see.

Humans have some vestigial structures, too: our wisdom teeth remain from a time when we had much larger jaws. Jaws have shrunk over time ever since we started cooking our meat, which makes it easier to chew, and inadvertently crowded out our back teeth. When you get goosebumps, those are your Arrector Pili muscles activating. Back when we had fur instead of just some thin hairs on our bodies, this response fluffed us up against the cold and made us look bigger to predators. All sorts of mammals do this; we just don't have the hair for it.

Likewise, cultural norms and moral ideas may also be vestigial. Aversion to pork, or even meat, in many cultures is not quite as adaptive today as it probably was pre-sanitation, when foodborne illness and other pathogens were harder to detect. However you feel about it ethically, pork is generally safe to eat these days, though pigs can still be a vector of disease from time to time. Same goes for dietary restrictions about shellfish, even though some people are allergic.

Supposedly, the handshake became a popular method of greeting because it showed the other person that one was not carrying a sword or dagger, so long as we were right-handed and not one of those sinister lefties (pun intended). Presumably the handshake has declined in usefulness since our weaponry changed from swords to firearms, and yet it persists, despite its effectiveness in spreading disease.

De Beauvoir argues that patriarchal systems, in which men have far more social power than women, were also made obsolete by industrialization. Physical strength no longer matters when machines do all the heavy lifting. De Beauvoir shows that societies' treatment of women as second-class citizens no longer retains any shred of logic, and yet the vestigial patriarchy still persists in the modern world.[57]

Like the idea of cultural adaptation itself, it is difficult to prove "vestigial cultural ideas" like these, to show that there is no remaining purpose for something, because it's always possible to conjure up some alternative explanation. In fact, we are probably seeing a lot more of what biologists call "co-option" in culture than "vestigial structures."

"Co-option" occurs when an organism uses one of its features or characteristics for a different function than it was originally selected for. Feathers are a great example of this. We now know that dinosaurs, the not-so-distant ancestors of modern birds, had feathers. In fact, they had these feathers long

---

[57] By the way, I am not asserting that the only reason we *should* treat people equally is because technology has allowed it. I am just sharing one possible reason for *how* these systems of inequality developed in the first place, and how technology is making it easier for them to change. Men, being physically stronger than women, had an advantage when it came to certain kinds of manual labor and warfare. Technological advancement has made that physical difference less and less important in the real world. In fact, women may now be outperforming men in the labor force.

before flying dinosaurs were around. Initially, they probably worked similarly to the way fur does for mammals: feathers kept dinosaurs warm, and could be used in mating or other kinds of social displays. It was only later that feathers were used to help hold early birds aloft in the air — natural selection acted on pre-existing structures and repurposed them.

If we look deep into our cultures, we often find similar co-options. For a religious example, take Christmas. The holiday initially existed as Saturnalia for the Romans, celebrating the winter solstice. In fact, many pagan festivals took place during the solstice. When Christianity became dominant across Europe, these festivals were modified to suit the needs of the new religion, that is, to celebrate the birth of its key figure.

There may be plenty of behavioral co-option in our cultures. Lots of animals, like birds and insects, give "nuptial gifts," a kind of food presented to a potential mate. By giving these gifts, an animal shows both its success as a hunter or forager and its ability to provide for young. It's no coincidence that males are almost always the ones giving nuptial gifts.

In human cultures, a common practice is to give gifts of food on dates or during romantic holidays like Valentine's Day, like a box of chocolates or an expensive restaurant meal. In fact, food-related gift giving and marital courtship are closely intertwined in dozens of cultures around the world and throughout history, possibly even going back to our pre-human ancestors. In the developed world, giving someone a box of chocolates doesn't necessarily mean you're interested in marriage, or even sex. It's more a way of conveying affection. We have co-opted the cultural behavior of generations past into a newer, modern form that fits present-day social behavioral norms.

The Constitution of the United States is another excellent example of cultural co-option. It has been amended over the years, but moral judgments codified by a small group of people hundreds of years ago — in vastly different circumstances — set the precedent for ongoing social and political behavior in the United States. How does the first amendment apply to anonymous, online forums or company-controlled social media websites? The fourth amendment protects against unreasonable searches and seizures, but what about digital possessions?

The second amendment of the Constitution was written partly in order to ensure that Americans could form militias if the British ever returned — which they did. It was also written at a time when citizens owned slaves and wanted to keep them in line. Twisted as it was, you can see why the wealthy, landowning white men wanted a guarantee of access to weapons. The question now is: how is that second amendment reframed in a world of mass shootings and automatic assault rifles? "Arms," in the late 18th and early 19th centuries, referred to muskets and flintlock pistols, weapons that were inaccurate and took half a minute to reload if you were well-trained. The founding fathers could not have contemplated the modern machine gun when the Constitution was written, and yet the second amendment is almost always invoked or implicated in conversations about gun control — an existing structure repurposed for new circumstances. That's co-option.

Religious texts offer more examples. Most religious documents are far older than the U.S. Constitution, and yet are still relevant for modern, technology-driven problems. Abortion is a highly contentious issue because technology has made it more accessible than in the past. One question at the core of this issue is: when does a human life begin?

Many conservative Christians believe that life begins at conception, the moment an egg is fertilized. Co-option of Biblical language is sometimes used to discuss abortion, though there is no explicit reference to abortion in the Bible. In a letter to the editor published in *The Sentinel*, entitled "Bible proves that life begins at conception," a reader commented, "And in Genesis 2:24 we are told 'Therefore man shall leave his father and mother and be joined to his wife; and they shall become one flesh.' That is the DNA of the male joining with the DNA of the female to become a new and wholly unique human being. God didn't say, 'Here's the mold for baby boys and here's the mold for baby girls ...'"

The reader then cites Jeremiah 1:4-5, which reads "Then the word of the Lord came to me saying 'Before you were born I knew you; I ordained you a prophet to the nations." The reader argues that this passage, along with a few other passages, demonstrates that human life even *predates* conception, as God has anticipated every human yet to be born. The reader concludes that "as to your 'biology lesson,' it is correct, but of no value to this discourse since, as Christians, we believe life begins at the moment of conception."

Intriguing: first the author takes the idea of DNA and retroactively applies it to a verse written long before humans had any idea what DNA was, effectively co-opting it. Next, the author co-opts the idea of predestination to show that human life begins at conception, and that each life is sacred. Remember, humans did not know about sperm and egg cells when the Bible was written. Besides the assumption that we have souls that occupy a zygote at the moment of conception, it's a wonderfully co-opted argument.

Interestingly, some Orthodox Jewish scholars also believe that a soul enters a body at conception, but that does not make the body "somebody." Instead, there are three stages. First, the *maya b'alma*, which is little more than a bunch of watery proteins, then an in-between stage starting 40 days in, then follows the *nefesh*, a fully realized person with a soul at the moment of birth. This translates into a markedly different view on abortion for Orthodox Jews, compared with Evangelical Christians, though obviously neither side holds a singular opinion. The point is that the moral grounds you start with can have a huge impact on how you address issues surrounding the technology and morality of the future.

Our "new" moral and cultural ideas about advances in technology don't come from out of the blue; they are usually deeply rooted in the precedent of generations past and are then co-opted to suit the times. Sometimes this works out great. Other times it doesn't work out at all. It's still encouraging to see the creative ways people are taking the texts and moral quandaries of old and using them to tackle the issues of a rapidly changing world today. For instance, it's no coincidence that environmentalism grew rapidly as a philosophy and movement in the mid-1800s, right when industrialization was starting to have a profound impact on the natural world around us. The race to conserve life, including human life, has only grown more dire over the years. Nowadays some Christian denominations re-interpret Genesis's "Be fruitful and multiply, and fill the earth, and subdue it; and rule over the fish of the sea and over the birds of the sky and over every living thing that moves on the earth" as a call for environmental stewardship. If God made humans in His image, and God is a benevolent ruler who cares for us, then we should likewise care for the world God put us in charge of. I greatly prefer this interpretation to one that praises the goodness of wantonly taking from the Earth.

The advancement of technology marches onward. It is likely that the future will be filled with moral and ethical challenges that we can't even think of now. The good news is that we do not have to know the *exact* nature of these challenges — just as we are doing now, people will re-interpret and co-opt today's moral precedents as they evolve into the moral beliefs of tomorrow. Evolution and adaptation are by their nature imperfect systems, and cultural or moral adaptation are bound to be even more imperfect when it comes to solving the crises of the present and future.

What moral precedent will we set for genetic modification? Environmental stewardship? Artificial intelligence? Space colonization? Interacting with aliens? If and when people have to deal with these future moral challenges, we will look back at the way we behaved in the past — what's now the here and now — as a source of guidance. What kind of standards do we want to set for them? What stories do we want to leave behind in our "cultural DNA"?

We have the fortune or misfortune of being born in a highly transitional time. We have the ability to manipulate genes, but we haven't perfected it. We can explore space, but not beyond our solar system. We can build artificial intelligence, but not machine sentience. Still, look at the progress we've made. Robotic machinery, modern medicine, flight, space travel, gene editing ... all of these technologies developed within the last century or so. Life on Earth is 3.6 billion years old, and just 10,000 years ago, we were still figuring out how to plant crops! Our species is still in its infancy, and we have so much potential. Yet massive existential threats like climate change and nuclear war still loom over us every day.

It's anyone's guess whether we will be able to resolve or at least mitigate these issues. For many of these threats — especially climate change — there is a time limit. Our actions over the coming fifty years or so will determine the direction of the next three hundred. As we continue to wage battles of morality and ideology, it is important to understand this: how we deal with the climate crisis today will be looked back on and judged for millennia to come.

23. Reconciling Morality and Evolutionary Biology

> "Personal annihilation sinks in my mind into insignificance compared with the idea or rather I presume certainty of the sun some day cooling and we all freezing. To think of the progress of millions of years, with every continent swarming with good and enlightened men, all ending in this ... Sic transit gloria mundi."
>
> — Charles Darwin, *Letter to J.D. Hooker*

His early years of science took him all across the world, across vast oceans and towering mountains. His final years of science take place in a garden.

The elderly Darwin runs a tight ship at Down House, but it is a loving home. His sons and daughters are growing up, becoming scientists, scholars, writers, and thinkers. Darwin, now both a father and a grandfather, is immensely proud. He writes papers on all manner of plants – including carnivorous ones – with Francis. Their garden blooms with color and variety. Daily life gets harder with age. But though Darwin has slowed down, he never stops. He tunnels through the loamy soil of science and writes one last book about his latest obsession: worms.

His house continues to be a laboratory unto itself. Darwin and family poke and prod around the garden, spades in hand. Rainy days are best. They carefully excavate and place the earthworms into pots, watching as they dig burrows and plug them with leaves. They get up in the middle of the night to spy on the worms' activities and measure rates of soil turnover. Darwin even asks his son and wife to play music near the worms to test if they can hear the sound. He deduces that they don't exactly hear, but detect vibrations through a solid medium. The groundskeeping staff at Down House worry for their employer when he spends minutes at a time staring at a leaf or squinting at the ground. Even in old age, that childlike curiosity about the natural world– and the patience to simply sit and observe– has not faded.

The goings-on of the worms beneath the soil, as they till the earth and transform dead vegetable matter into fertilizer, are a quiet reminder to Darwin about his own place in the cycle. He is short on time. Darwin doubts that the general public will care much for a book on worms, but asks that it be published just the same.

He's shocked when it ends up being a bestseller. The worm-wrangler's postage is flooded with fanmail from fellow gardening enthusiasts. His contemporaries praise the book, applauding the way Darwin showed the value of even a small and lowly creature. American minister Moncure Conway writes an entire essay based on "the new epic, which Darwin has gathered and gifted... this new song will reveal the worm as the reverse of the destroyer. The worm is the resurrection and the life... Things hard and useless for the life of Nature– , stones, ruins– it buries, but converts organic remains into mould, and brings them to earth's surface, where they may bloom again and breathe in fragrance and happy carols."

There is no major controversy in this final book, no political upheaval. There is only the simple reassurance that even small beings– perhaps small beings especially– are key drivers of life as we know it. In the end, it is the collective action of many small beings, all working together, that allows the garden of life to thrive.

At the end of his life, Darwin looks at death without fear. He looks back with contentment and gratitude for a lifetime of discovery and a loving family. Charles sees at the end that he, like each of us, is both the gardener and the garden. He is both a shaper of Nature and a part of Nature. Even when he passes on, that will not change.

His health rapidly declines in the beginning of 1882, just before his seventy-third birthday. By mid-April, he is on his deathbed. The connection to the natural world and the people who inhabit it– these are what ultimately gave him purpose. These connections are what defined his life, and what will define his legacy. Bound for the next adventure, Darwin's last words are those of a man who experienced the fullness of life and at long last can call his life's work complete:

"I am not in the least afraid to die."

///

Humans have always dealt with chaos and uncertainty, but our global ecological crisis is on an unprecedented scale. It's hard to visualize, let alone tackle. Moral adaptation in modern humans is therefore necessary, and with it, a paradigm shift in how we view the world.

All the battles and struggles and moral challenges that come with being alive are variations on a single theme that has persisted since the beginning of time: the struggle of organization against entropy. In scientific terms, "entropy" relates to disorder, spreading out, and cooling down. According to the second law of thermodynamics (discovered late in Darwin's life), the total amount of chaos or entropy in a system can either stay the same or increase. Things can only get more chaotic, or at best, stay as chaotic as they were before. As quoted at the start of the chapter, Darwin himself realized the troubling implications of this law, though scientists at the time did not understand nuclear energy and thought the sun would slowly cool over millions of years instead of getting massive in a few billion.

We now know the universe is expanding, and as it does, galaxy clusters grow further from each other and the average temperature of outer space decreases. Things tend to fall apart if energy is not constantly applied to keep them together — that's why my room can only get messier unless I actively do something about it. Stars, spheres of hydrogen and helium organized by gravity and nuclear bonds, disperse and radiate their energy into space over time. Even some of the most basic of particles, like protons, decay.

And yet on Earth, we bear witness to a wondrous aspect of our universe: life. Of every entity we've observed in the cosmos, life is the most diametrically opposed to going with the flow of entropy. The organic molecules we are made of hold together over time because life uses an elaborate system of organization that, time and time again, figures out ways to reproduce and persist. Each time a living organism survives and reproduces, a legion of organic molecules collectively figures out a way to keep itself together, while pitted against laws of physics that pull everything toward disorder. Life on Earth has weathered 3.6 billion years of meteor impacts, volcanic eruptions, climactic upheaval, and all manner of catastrophes. Watching organisms on Earth grow and adapt, in spite of life's many disasters, is awe-inspiring. The same goes for people. Every human family has a near-death story, a tale of hardship that an ancestor ultimately overcame. Life, in whatever form it takes, is a never-ending triumph.

Humans are extraordinary. The evolution of an animal as cognitively powerful as *Homo sapiens* took billions of years. Once humans *did* evolve, they almost immediately faced extinction via ice age. By the skin of their teeth, our ancestors weathered every disease, predator, and resource depletion the universe threw at them, and from those humble beginnings we went on to sculpt the entire planet to meet our needs. We should remind ourselves that what human civilization is attempting now — organizing more than eight billion humans into co-habitable groups — is a major challenge.

History is a great teacher; it shows us just how quickly civilizations can collapse under the right circumstances. Anyone who has lived during a time of societal upheaval will understand how fragile communities are. As an example: money, the way we represent the flow of resources, exists entirely on collective faith. If everyone stops believing that a dollar has value, then it is just a piece of paper. If everyone stops believing Rome is an empire, then it ceases to be one. We expend colossal amounts of energy reinforcing our beliefs, which keep ourselves, our families, and our societies intact. Our ability to reach a moral consensus and act collectively plays a massive role in keeping us united in purpose.

Unfortunately, we can't shake the vestiges of our past. Old biological and cultural biases affect how we view and interact with others. Humans are hardwired to be simultaneously cooperative and competitive, and we take cognitive shortcuts. Racial prejudice, violence, and selfishness pull us back toward disorder, making the achievement of "progress" — happier, more equal societies — an uphill battle.

There is a silver lining, though. Our tendencies toward cooperation and competition can be positively channeled, especially when people unite against a common enemy. Historically, this has involved people of one religion uniting against people of another, or else a battle between competing ideologies like capitalism and communism. It doesn't have to be that way. Barring an alien invasion, humanity can unite against another common enemy, extinction, and toward a common goal, the preservation of life.

We already ascribe great moral value to life. Our religious and cultural texts are filled with divine mandates and spiritual stories that sanctify both humans and the plants and animals that surround them. Darwin himself wrote that "disinterested love for all living creatures" was "the most noble attribute of man." Facing a mass extinction brought about by human-driven climate change, invasive species, overharvesting of resources, and widespread habitat loss, the sanctity of life will only become more central in our moral and ethical struggles. As it should be — given the consequences of ecological collapse, the sanctification of life is a critical moral value.

The preservation of life on Earth will be the single greatest challenge of the 21st century. The lives of billions of humans and countless billions more organisms are on the line, along with centuries of future technological and societal innovation. People in countries like the United States increasingly report a loss of direction, meaning, and hope. Is collective environmental action exactly what humans need? Can we reignite our sense of purpose?

Even now, the fight to preserve a recognizable Earth in the 21st century is gaining momentum. People around the globe are undertaking huge cooperative stewardship projects to try and counter pollution, extinction, and climate change. India, Pakistan, China, and Ethiopia have started massive reforestation projects, collectively planting billions of trees in

Above: Apollo astronaut Bill Anders' famous photo "Earthrise." Photo by NASA and Bill Anders in the public domain. Accessed via Wikimedia Commons.

the process. Conservation groups are investing millions into replanting in the Amazon rainforest. A multinational effort sponsored by the United Nations, known as the "Great Green Wall," is working to stop desertification in Africa. By creating a massive swath of forest running through the heart of the continent, the project will hopefully stop desertification in the region as the land keeps heating up. The Great Green Wall may be the largest ecosystem restoration project in the world, aiming to cover over 100 million hectares by 2030. Though replanting is not a magic cure for stopping the effects of climate change, it's reassuring to see thousands of people working toward a common environmental goal. If the Green Wall is completed, it will stand as the largest living structure on Earth, three times the size of the Great Barrier Reef, and a testament to what collective human action can accomplish.

We've already seen victories in animal conservation, too. Bald eagles teetered on the brink of extinction in the 1960s, with only 500 or so nesting pairs left. Today, thanks to conservation efforts, breeding programs, and the U.S. working to ban the egg-destroying pesticide DDT, the bald eagle has been off the endangered species list for well over a decade. Grey wolves, down to only 300 or so individuals just half a century ago, now number over 6,000. Human activity devastates sea turtle populations through harvesting, habitat loss, and light pollution, but thanks to local communities' efforts to protect beaches for nesting turtles, sea turtle populations have skyrocketed in some areas. It's not been easy, but people are starting to make a positive difference.

This is not our first extinction rodeo. We drove countless species of megafauna to extinction in our mammoth-hunting days during the Pleistocene. For a while, it seemed like whales were next, but thanks to worldwide restrictions on whale hunting, populations of critically endangered whales like the blue whale are finally recovering. In the North Atlantic, the previously overhunted humpback whale is approaching estimated pre-whaling populations. "Save the Whales" became a rallying cry in the late 20th century, and politicians listened. Today, huge swaths of the oceans are designated as whale sanctuaries, and even though people still cheat and illegally hunt whales, we're on the right path to saving them from extinction.

A powerful tool for humanity, as it has been for hundreds of generations of our ancestors, is our empathy for other living beings. In recent decades, that empathy has allowed some of us to extend our in-group to include the wildlife around us — even to the biosphere as a whole. Some of us are so empathetic to other organisms, we've become vegetarians. Mr. Rogers famously said he "couldn't eat anything that had a mother." It won't be Eden, but if we continue to make progress and undertake collective action, the biosphere — and by extension the human species — will make it through this ecological crisis intact. Above all else, no matter how rough the coming centuries get, we cannot — will not — give up the struggle. Empathy, resolve, and patience are more useful than fatalism.

In the short term — for us, our children, and our grandchildren — our primary goal is to prevent the collapse of the biosphere. In the long term, though, we should not stop with climate change. If we value life on Earth, another threat looms: the permanent end of life on Earth. "Some say the world will end in fire," wrote the great poet Robert Frost. "Some say in ice." The ice people were wrong. The world is going to end in fire.

The sun sustains life on Earth via the energy it radiates, and it has been getting brighter throughout its lifetime. The sun will continue to brighten as time passes, with no regard for the planets it will eventually scorch. Hundreds of millions of years from now, the sun will become so bright that the Earth's oceans will evaporate, along with all other large sources of liquid water on the planet. Without plentiful water, it's doubtful that life will be able to survive in any more than the most basic, single-celled forms. Even that won't last forever — when the sun enters the next stage of its life as a red giant, billions of years from now, the Earth will be completely consumed, and all living things on the planet utterly vaporized. Life on Earth is ultimately doomed — or it would be, if not for one ambitious, self-congratulatory primate species.

Humans, for all their faults, have the potential to change everything. We are the only species on Earth with the potential to travel off this planet and onto others. Sure, our space travel is still in its infancy, but if the past couple hundred years of technological progress are any indication, it really won't be long before we build ships that can take us to other stars, potentially even other galaxies given time. If that sounds far-fetched, note that, a few decades before the Wright brothers flew, journalists wrote that the invention of powered flight would take thousands of years. Space-faring humans are no wild fantasy.

Maybe another species could evolve after us with the potential to go to the stars, but the probability of that happening is low. As far as we know, it took billions of years of evolution to produce one tiny branch of life with the cognitive ability to build spaceships. If we wipe ourselves out, it's unlikely that another spaceship-building species will appear in time. In other words, humanity might be the one shot that life on Earth has at escaping its own end.

Our sun is a medium-sized star with a moderately long lifespan of 10 billion years. The second half of its life will be spent as a red giant, once it begins to run out of hydrogen to fuse and begins fusing helium. Larger stars burn through their hydrogen reserves much faster than smaller stars, so the smaller the star, the exponentially longer its lifespan. A red star one-tenth the mass of our sun, for instance, will live for somewhere around two trillion years without ever becoming a red giant. That's 147 times the current age of the *entire universe.* Imagine how many species could evolve in that timeframe —

how many potential intelligent life forms. While living with a red sun doubtlessly has challenges, what if humans were to travel to a red star, terraform an uninhabited planet, and introduce even *bacterial* life there? We would beat back entropy for trillions of years, extending the timeline of Earth's life a thousandfold. For all of humanity's "sins" — causing extinctions and environmental devastation — this single act might outweigh all of the damage we've caused many times over.

Agriculture changed the world. "Bioculture" could change the universe. Humanity could, if it so desired, eventually become a prolific cosmic gardener, taking life and cultivating it all across the galaxy. We could grow and adapt alongside the species we bring to uninhabited worlds, taking on whatever physical or social challenges the wide universe throws our way. We will never run out of things to do or discover. This is, at the least, a pleasant fantasy.

But let's get back to Earth. I'm not suggesting that our becoming spacefaring stewards of life is a realistic near-term goal, but it's a worthwhile long-term goal to strive for while defeating the four horsemen of "disease, poverty, unhappiness, and violence", as Yuval Harari puts it, in the meantime. It's not the end result that will make our futuristic environmental stewardship useful, it's the steps required just to get there.[58] Resolving the challenges of the present day will require collective action that puts the moral value of life over short-term wealth. Our decades of ideological gridlock, warring over

---

[58] By the way, it's not like environmental stewardship is a new idea. As we discussed earlier, Native Americans have been caring for large wildlife areas for thousands of years.

moral differences, and science-versus-faith debates have to end if we're to make it through the next couple of centuries.

Avoiding ecological disaster requires radical changes in policy through economic incentives, improvements in education, and a vast mobilization of the workforce. If we're going to aim for the massive undertaking of conserving the entire planet, we need everyone on board. My grandma used to say, "Idle hands do the devil's work," a saying actually backed by statistical evidence. We need to help people find purpose.

Cultural co-option could go a long way in making a better world. For instance, the United States has a highly militaristic culture. American fighter jets fly overhead during sporting events, veterans are routinely singled out as heroes,[59] and nearly a trillion dollars are spent each year on the U.S. military, more than the next ten countries combined.

On the surface, militaries exist to confront *all* threats, foreign and domestic. Could we co-opt the existing enthusiasm for the military to combat existential threats? We needn't get rid of the Army, Navy, and Air Force, just reprioritize the trillion dollars of funding. Instead of bombing people from developing nations, use some of the military budget to replant and guard rain forests and wildlife sanctuaries. Instead of building more bases, oversee the construction and repair of key strategic infrastructure to protect coastal cities amidst rising sea levels. The military could reshape America's transportation systems with innovations like high-speed railways, just as it built the interstate highway system in the 1950s. Instead of new fighter jets and tanks, the military could help fund wind turbines and solar panels, ending our dependence on oil.

---

[59] Despite this, veterans in the United States have high rates of poverty, depression, and suicide. You can't live off of standing ovations.

Climate change is a threat to the security of the world, and militaries know this already. Why not start officially treating it as such? Why don't we wage war on climate change and mass extinction?

Humans change the world quickly, but big societal change still unfolds slowly compared with individual lifetimes, let alone evolution. In the case of climate change, slow and steady does not in fact win the race. In order to change quickly, our *existing* cultures and philosophies must adapt.

Religion unifies people, motivating them to collective action — think of all the art, architecture, and warfare made from religious motivations. Religion also has positive effects on mental health. It would be an excellent idea to frame the existential, difficult-to-grasp scope of climate change in a religious and/or spiritual context, supported by moral perspectives from religions around the world. The threat of extinction inspires spiritual introspection, cohesion, and purpose which in turn helps many of us overcome hardships and deal with the constant stress of climate change.

What if climate change were not just an ecological phenomenon, but a challenge that we as stewards of the planet were spiritually motivated to overcome? Not only could framing climate change through a spiritual lens help focus and motivate some groups of people, it could also help foster a worldwide sense of community, making the daunting prospect of tackling climate change feel more manageable. Social change requires "people power," and religions motivate wellsprings of it.

For too long, "big thinkers" on either side of the "evolution vs. creationism" argument have tried to pit science against religion. Whatever your stance on the merits of religion, we don't have time to argue over minutiae. It shouldn't matter if you think Adam and Eve — or Mitochondrial Eve — were real, or if God made us directly, or if the Earth rests on the back of a giant turtle. No matter what you believe, the giant patch of plastic in the ocean floats right in front of us. Smoke and smog turn the sky blood-orange. Animal species are vanishing. Rainforests are burning. We don't need a *debate* that there's a problem, do we? Then let's shift our attention to what we agree on, what we can do, how we can cooperate.

Human societies are naturally competitive. Let's tap into that and make tackling climate change a friendly competition, like the World's Fair or the Olympics.[60] Along with technological collaboration, countries and companies alike could compete, producing new innovations, projects, or leaders in the midst of the crisis. Host a climate protection awards show with some grand prizes, and frame it like *Eurovision*. With the right P.R. team and a good deal of money, anything's possible.

Cultural and moral adaptations work because as much as technologies change, people are still people. Our brains are pretty much identical to those of our ancient ancestors, and so are a lot of our social interactions. Some of our behaviors are so old that they predate the human species itself. We can take advantage of that, too. Our knowledge of human behavior and cognitive biases could be turned to our adaptive advantage. Rather than using neuroscience for market-driven brainwashing and smartphone addiction, we could figure out a way to communicate effectively with people about important

[60] The video game *Sid Meier's Civilization VI* has a feature whereby different civilizations can compete with one another to mitigate climate change and win "diplomatic victory points." A fun concept. Hopefully it doesn't end up being as fantastical as the game's giant death robots...

issues. Humans remain distracted and divided by long-standing prejudices and feuds, the evolutionary results of in-group bias. The pain we inflict on one another keeps species-wide cooperation out of reach. Science and technology should unite not to polarize humans, but to promote better communication and mutual understanding. Policies concerning education, healthcare, crime, and more could all be improved by a behaviorally informed approach.

Empathy is a powerful tool, and improvements in communication will be essential to create consensus and confront common threats. We need all people to view all other people as part of their kin, part of their team. No matter what happens with our responses to issues like climate change, artificial intelligence, genetic manipulation, et cetera, it is going to be a brave new world. Science and technology have no moral stake — *we* do. The question is whether we will use our newfound knowledge to dominate and destroy, or to unify and uplift. It's a race against time, and self-destructive behavior cannot be allowed to win out.

Here, I am reminded of a poignant example from the Dutch author Rutger Bregman's excellent book, *Humankind, A Hopeful History*. In it, Bregman recounts striking behavioral changes that occurred during The Blitz– Germany's bombing of London during World War II. The Blitz was meant to break the spirit of the Londoners– Hitler believed that after many days of strife under constant bombardment, the British would turn on one another. The bonds connecting the British cultural group were expected to weaken under continual duress. When they fractured, the Germans would take London.

But instead, the opposite happened. Rather than turning hostile, the people of London actually became *nicer* to one another during the Blitz. Rates of depression and suicide were lower. People reported a boost in mood. The hardship that people collectively endured during the Blitz brought out their strength of character, causing people to be *more* willing to help and support their neighbors. One journalist wrote that "The courage, humor, and kindliness of ordinary people continue to be astonishing under conditions that possess many of the features of the nightmare." This upswing in kindness wasn't unique to Britain– all around the world, people have banded together more closely during times of collective trauma like bombings or natural disasters.

Rebecca Solnit, author of the book *Paradise Built in Hell*, describes scenario after scenario where human kindness shone in the face of catastrophe. Despite media coverage to the contrary, Hurricane Katrina brought volunteer rescuers out in droves. Everyday people headed to New Orleans in boats to look for survivors and banded together to scavenge up medical supplies. After Hurricane Sandy hit New York, crime rates plummeted in areas where looting was expected to run rampant. Whether it's an earthquake in Mexico, a tsunami in Japan, or a flash flood in Iran, we time and time again prove that humans are inherently brave, and inherently kind.

Our evolutionary history as primates who banded together during a near-extinction event changed us. Compassion forged our species. Ordinary people do not turn on one another when the chips are down. They lift each other up. The science is clear: when humans undergo extreme hardship, we become the best versions of ourselves. If we can channel that same feeling of goodwill under the deadly effects of climate change– if our species can recognize its universal kinship in the face of disaster– perhaps *Homo sapiens* can still emerge from this crisis kinder, more resilient, and changed for the better.

24. What's the Problem with Evil?

> "Our descent, then, is the origin of our evil passions!! — The Devil under form of Baboon is our grandfather!"
>
> — Charles Darwin, Personal Notes

> "But I am very poorly today & very stupid & I hate everybody & everything. One lives only to make blunders."
>
> — Charles Darwin, *Letter to Charles Lyell*

Changing the world for the better means confronting "evil." We often give voice to a minority of people who don't make the world a better place, who seem to be trying hard to make it worse. The issue of how much "evil" there is in the world inspired me to write this book. I'm an optimist by nature. I believe that *most* people are trying to do the right thing.

The classic response to this way of thinking says, "Well, what about the Nazis? They were evil, and the majority of Germans went along with them." I don't understand why this bunch of white, nationalist, genocidal Germans from the mid 20th century always seems to substitute for the entire human species in these conversations, but fine. There's no doubt that evil, or something "profoundly immoral and wicked," is pervasive in the world today. I use "evil" loosely here, referring to immoral behavior that the vast majority of civilizations try to avoid. Like moral good, its opposite is a relative concept, ever shifting with the times. What's perceived as evil today may have little in common with what was perceived as evil in centuries past. And also like moral good, just because evil is something we collectively conjure does not make it illusory. As moral acts are adaptive in the proper context, immoral acts can be *maladaptive*, and that is precisely why humans evolved to be repulsed by such acts, and why we have systems of punishments in place for offenders. It is important to recognize how and when immorality occurs, which, much like morality, is attributable to three main causes: ingroup vs. outgroup competition, mental conditioning, and vestigial (outdated) animal behavior.

Let's start with the first cause, "us vs. them," which is probably the most common source of evil on a systemic scale. Darwin writes:

> A tribe including many members who, from possessing in a high degree the spirit of patriotism, fidelity, obedience, courage, and sympathy, were always ready to aid one another, and to sacrifice themselves for the common good, would be victorious over most other tribes.

This is no doubt true, but sometimes the "common good" of one group can be the common enemy of another. Humans are sometimes brought together as a group, ironically, by their mutual despising of another group. Taken too far, pride in one's own habits and customs can turn into revulsion of another's, even if they are harmless. It is no coincidence that nationalism and xenophobia go hand in hand. Focusing on details that appear to distinguish one group from another often leads to seeing those details as points of superiority or inferiority.

Japan is a remarkably adaptive civilization, and despite large cultural differences, is close with many Western Nations.

It is also extremely nationalistic, though not *as* nationalistic as it was circa World War II. Japanese citizens were brought up as staunch collectivists, raised to be devoutly loyal to emperor and country. They were also raised to despise neighboring nations, especially China. At the beginning of World War II, Japan was eager to expand its territory and acquire resources from its neighbors. This combination of self-supremacy, vitriolic xenophobia, and an extreme cultural aversion to the concept of surrender created the perfect storm in Nanking.

What the Japanese army did in Nanking is considered to be among the most brutal, evil acts perpetrated during World War II. Chinese civilians were beheaded, tortured, raped, mutilated and set on fire en masse by Japanese soldiers. I do not need to describe what happened in any greater detail, but the event led to some 300,000 deaths over the course of only a few months of occupation. It is not just the number, as the late historian Iris Chang points out in her book, *The Rape of Nanking*, but the manner in which these lives were taken that is truly horrifying. "So sickening was the spectacle," Chang writes, "that even Nazis in the city were horrified, one proclaiming the massacre to be the work of bestial machinery." One shudders at the thought of actions so terrible that even *Nazis* balked.

War, genocide, slavery, and other terrible acts perpetrated by human groups rely heavily on the dehumanization of the "other." Japanese men and women were raised in an intense social hierarchy, amplified by militarization and the belief that the Japanese were a super race, destined to rule over other races in the "Eastern Co-Prosperity Sphere." Here's a 1942 clipping from a Japanese newspaper about foreign soldiers who surrender:[61]

> They surrender only when they have sacrifice all lives in their power excluding their owns. When they finally surrender themselves, then only to save their own lives. They have shown in all previous battles that they are totally egoistic. With this behaviour they have reached that we cannot treat them as normal prisoners. They have broken Gods law, and therefore God will punish them with defeat. However, any mercy would prolong the war. The Japanese soldier fights in a holy war, in which there is no place for think over and uncertainty. Criminals must be annihilated.

This passage is telling, in terms of Japanese groupthink at the time. Surrender was impermissible to Japanese citizens and soldiers, and those who did not share this cultural idea were deemed subhuman, as were Western POWs and Chinese prisoners. "Mercy" and "uncertainty" in the face of violence are often sacrificed to the perceived needs of war, effectively justifying evil behavior. Across the sea, the United States was likewise willing to bypass its Constitution, forcibly interning thousands of Japanese-American citizens in camps, under the assumption that there were spies among them. The government deemed this act "justified" in war. Historically, when groups of people ignore their own morality and act immorally, they almost always look for a justification.

The Nazis exemplified this need to justify. Their racial pseudoscience and radical nationalism dehumanized Jews,

---

[61] I have left the paper's typos in.

people of color, and other non-Aryans, to the point that their individual lives were deemed worse than worthless. Their outright elimination was framed as a tough but ultimately necessary "final solution." Some say that lack of religion drove the Nazis to behave in this way, but that's not true. Hitler himself said, "I believe today that I am acting in the sense of the Almighty Creator. By warding off the Jews I am fighting for the Lord's work." The Nazis thought God was on their side, too.

Systemic immorality also allows people to exempt themselves from blame for the evil acts they commit. The phrase "just following orders" haunted the Nuremberg trials, where the full atrocities of the Holocaust were broadcast to the world. People were horrified by the emotional detachment of it all — "the banality of evil," as writer Hannah Arendt put it.

Systemic immorality is both pervasive and probably the most difficult to discourage. Perpetrators collectively decide that their actions are somehow justified, in spite of basic moral norms, because the people they torture or slaughter are "not fully human." Indoctrination and groupthink probably gave humans an adaptive edge early on, but it came with a heavy cost. Few things are more dangerous than an immoral group of people that believes it is doing good.

Systemic immorality is different from immorality that stems from mental illness, though the two can interact. Some immoral acts are traceable directly to abnormalities in the brain. Some people may simply lack the brain functionality for empathy, or the capacity to understand that what they are doing is wrong, due to developmental conditions.

In one example of a brain gone awry, an otherwise ordinary 40-year-old man seemingly developed abnormal social and sexual behaviors spontaneously, including the viewing of child pornography, which he had never had any interest in before. The day before the man was to be sentenced to a 12-step rehabilitation program, he visited the emergency room, complaining of a severe headache. According to the case report, after he continually made advances on female hospital staff, doctors noticed other bizarre behaviors, such as twitching, an unbalanced gait, lack of bladder control, and an inability to write clearly. Doctors suspected a brain tumor, and sure enough, they found one lodged right by the orbitofrontal cortex, the part of the brain that deals with the regulation of social behavior. After the tumor was removed, the man's gait returned to normal, and so did his sexual and social behaviors. When the odd behavior resumed sometime later, doctors discovered that the tumor had returned. The man's deviant behaviors ceased altogether after a second surgery.

According to the man's psychiatric evaluation, he knew that his actions were wrong, but he simply could not overcome the overwhelming urges in his brain. The man's behavior *was* profoundly immoral, but its root cause wasn't groupthink or the way he was raised; his brain was literally broken. There's some cause for optimism there. As medicine improves, harmful behaviors caused by mental illnesses may be more easily cured or managed.

Apart from mental illness, immoral acts may also occur as a result of not-so-voluntary biological functions. Take the fight or flight response. This instinct is your body's panic button when your brain doesn't have time to think things over. In the heat of the moment, making that snap decision to stand and fight or turn and run can potentially save you and your entire lineage.

Sometimes, a heated argument can turn deadly at a moment's notice. Overwhelmed with emotion in a heated argument, one party may pick up a rock, and without thinking, strike the other person in the head. The killer is left in disbelief. He is not a murderer, after all, just an ordinary human being, or so he thought. Primal instincts, hardwired from when we had to fend off lions or tigers or bears, sometimes kick in and inspire us to violence before we even know what we were doing.

Scenarios like this — "crimes of passion" — do happen, but they seem to be more rare than the first two patterns of "evil" behavior. Given a deep enough understanding of the workings of the human brain, the concept of "free will" gets pretty shaky. This ambiguity has led neurobiologists like Sapolsky to suggest a reworking of our entire criminal justice system in light of our ever-developing understanding of the human mind. The idea that people can spontaneously commit evil acts even when they *know* it's wrong is a sobering thought, but an important one to consider. Our legal systems were not designed with an understanding of neurobiology. Perhaps it is time that they adapt as well.

Sources of "evil" are also not mutually exclusive. Our fight-or-flight instinct can be affected by internalized racism. Mental illness can pair with socialization to create some truly awful behaviors and ideas. In pursuing an understanding of evil, we should not act to *excuse* these behaviors, but rather to understand where they come from– to limit their presence.

And of course, there are moral gray areas as well. Theft is generally considered an immoral act. But what if Jean Valjean stole a loaf of bread to feed his family? What if a thief shook down a helpless older woman for her purse, but needed the money to fund his child's surgery? Who do we blame, the thief or the system that put him in a desperate position in the first place? Is it okay to kill if you know for a fact that, if you don't, your whole family will be murdered? When usually immoral acts are driven by desperation, I don't feel the need to describe them as evil.

It's also important to acknowledge that although people have done and continue to do terrible things, *most* humans are, by and large, kind, empathetic, and willing to help others. Our brains are hardwired to focus on the negative stimuli we see in the world, while the positive stuff can often go unnoticed. Michael Shermer writes:

> An asymmetry in our moral observations about what people are really like comes from the fact that we have a tendency to focus on extreme acts of immorality and ignore the fact that most of the time, most people are gracious, considerate, and benevolent. For every act of violence or deception that appears on the nightly news, there are 10,000 acts of kindness that go largely unnoticed. In fact, violence and deception make the news precisely because they are so unusual in our daily experience.

Thanks to Darwinian natural selection, most humans *feel good* when they help others. We want to be good people. A Gallup poll revealed that 44% of people worldwide said they had helped a complete *stranger* in the past month. Nearly a third of people surveyed donate money to charity. A fifth volunteer their time to a charitable organization. A Columbia University study found that people tend to underestimate the number of strangers willing to help them by as much as 50%. We're not nearly as bad as we think we are.

So long as there is an idea of a moral "good," an immoral "evil" is always going to exist in juxtaposition. Taoists are spot on with Yin and Yang — you need one for the other. Otherwise, like we discussed earlier, the universe would be a boring, stagnant place with no complex life, no cooperation, no reciprocal altruism, and no happiness.

Definitions of good and evil may change over time, but their mutual existence in the minds of humans will not. A defining feature of the industrial and technological revolutions has not been the inclination toward "evil," but rather the increased *capacity* for it. In the "old days" before the machine gun was invented, violent, mentally ill people could not get their hands on assault rifles. Racist totalitarian governments did not have the technology or resources to build gas chambers. Democratic governments did not have the technology or resources to incarcerate more than two million people. Countries could not deploy nuclear or biological weapons. Advancements in technology have enhanced our ability to do evil in proportion to our ability to do good. This makes it all the more vital that we get a better understanding of good, evil, and human behavior soon, because the consequences of failing to do so will be increasingly disastrous over time. We owe it to future generations to create a world where our worst impulses, remnants of an evolutionary history filled with struggle and competition, no longer dictate our behavior on the global scale.

25. The Voyage Ahead

> "From death, famine, rapine, and the concealed war of nature we can see that the highest good, which we can conceive, the creation of the higher animals has directly come."
>
> — Charles Darwin, *The Foundations of the Origin of Species*

Charles Darwin struggled greatly to comprehend a morally ambiguous universe. His quest for knowledge was accompanied by love, heartbreak, loss, joy, anger, and a constant examination of whether the law-giving God that he believed in was truly good. And yet, through it all, Darwin's work rocked the foundations of science and faith to their cores. Despite his suffering, Charles Darwin imagined a universe whose laws *could* lead to the development of moral beings, whose laws *did* create animals capable of loving one another as they loved themselves. Darwin noted in his autobiography that:

> Some writers indeed are so much impressed with the amount of suffering in the world, that they doubt, if we look to all sentient beings, whether there is more of misery or of happiness;- whether the world as a whole is a good or a bad one. According to my judgment happiness decidedly prevails.

The universe Darwin observed and experienced was one of strife and struggle, where organisms face pain, death, and destruction. Ultimately, it was also a world where good triumphed. The very mechanical laws of Darwin's universe worked in favor of the righteous, whether God was there or not.

From Charles Darwin's work, we've come a long way. Each of the social organisms we've studied has taught us something about ourselves: we're not moral beings by coincidence. We have good reasons to act in the ways that we do. Humans come from a line of social primates that learned to cooperate and band together in the face of near-extinction. Our ancestors used their empathy, intelligence, and culture to survive in a hazardous and unpredictable climate. Their social structures were held together by shared moral principles that encouraged cooperation *and* changed to suit their surroundings. They changed, and so can we.

All living things are locked in a never-ending conflict with a common antagonist: entropy. And yet, largely because of this conflict, our universe rewards mutualism and cooperation between different entities, making it possible for beings to band together and become more than what they can be on their own. The struggle for existence brings organisms together against a common foe, toward a common goal.

This is how entropy is resisted: unity. Atoms merge to become compounds. Amino acids merge to become the stuff of life. Prokaryotes merge to become eukaryotes. Cells merge to become multicellular organisms. Ants merge to become colonies. Trees and their symbiotic fungi merge to become forests. People merge to become civilizations.

The human species has a long journey ahead of it. Who knows what our technologies, combined with our ever-evolving moral compasses, will lead us to become?

As we reach the end of this book, I hope I've convinced you of, or at least helped you understand, a few key ideas. First,

Charles Darwin was a free-spirited man of many opinions, but he was *not* an atheist. He was a deist who struggled with the idea of the universe's benevolent design. His views have been highly edited and greatly simplified, contributing to the worldwide war between the spiritual and the scientific. The constant pitting of science against religion is unhealthy, unnecessary, and distracting from the bigger picture: saving ourselves.

We've seen that humans are not the only animals that act morally. Many animals — arguably even some plants — demonstrate fairness, cooperation, and self-sacrifice, especially as one climbs the cognitive ladder to animals like whales, great apes, corvids, and parrots. These animals may even possess a rudimentary form of what we call culture. It is not in spite of, but rather because of environmental and social hardship that moral mechanisms like culture, kin selection, reciprocity, and fairness exist at all.

As for humans, our cultural and moral ideas don't just spring into existence. They are tied to our underlying biology, and many of them were *selected for* over time. Cultures and morals evolve, adapting to new challenges posed by growing populations, new technologies, and changes in the climate. If we want to prosper in the coming centuries, to meet the human-driven changes and challenges that await us, we will have to adapt our cultures and our morals further, dedicating ourselves to an ideal higher than money and extending our empathy outward to life on Earth. If we succeed in maintaining ecological stability today, there is no limit to what we could eventually accomplish, or what we could eventually become. The universe is better with us in it. Let's keep it that way.

* * *

When the pandemic lifted enough for me to go safely back to fieldwork, the first thing I did was head to Yosemite. I was thrilled. At that point I'd been doing my animal behavior PhD in California for months but had never gotten to see much of the famed park system. I struggled to keep my eyes on the road during the drive up to our field site. From the winding highway, I looked out and saw a vast stretch of snow-capped mountains, green forests, arid desert, and a massive, sprawling salt lake. As a Midwesterner, well, I love the Great Lakes, but this was on another level. Yosemite was like a miniature planet, all the biomes just a few miles removed from one another — a wondrous ecological collage.

These days, I've been helping with some bird research near that wide swath of salt water called Mono Lake. Millions of migrating birds arrive there in gigantic flocks to rest and feed on brine shrimp that live in the water. These kinds of stopover sites are essential for the birds' survival. I'm trekking my way down near the shore one day, and I look out at the tufas, these strange, twisting towers of white calcium carbonate that form from the mixing of fresh underground spring water with the lake's alkaline water. There's a little rusted signpost across from me that reads "1960." That's where the water level was half a century ago. I would have been submerged.

In the 1940s, the city of Los Angeles began to divert water from Mono Lake's tributaries. By the '80s, the lake had fallen more than forty feet. The birds that relied on the lake, and its brine shrimp, were in serious trouble. The courts eventually realized the severity of the problem, in no small part thanks to the efforts of the passionate people on the Mono Lake Committee. A 1991 injunction greatly limited diversions from

Mono Lake, and the water has been rising ever so slowly ever since.

Thanks to just a few people acting on their moral beliefs in the value of nature, an entire ecosystem was saved. This story is a familiar one. Even now, millions are making their voices heard in the fight for conservation, not for wealth or fame, but from a great, unconditional love for their neighbors on the evolutionary tree of life.

Nature hasn't waited for policy changes. Now, where huge swaths of dry lakebed used to be, plants live instead. Fields of reedgrass grow out of the earth, and some adolescent trees dot the receded lakeshore. The iconic tufas are stranded inland, looking alien and out of place. I pull out my phone to take a picture.

And just then, a little feathered head pokes out of the inland tufa next to me. It's a violet green swallow, and he has been using the holes in the tufa as a nest. Very clever. He stares at me for a moment, preens his chest feathers, and flies off. I watch him swoop past the treeline. A minute later, he's back with a beakful of bugs.

For the nestlings, right? I think.

"PEEP, PEEP, PEEP!"

Yeah.

The swallow finishes feeding the kids and emerges once more onto the sunlit tufa, his chest swelling with exertion, and maybe pride. In the light, he glows a radiant emerald. He cocks his head and shoots me a side-eyed look as if to say, "Not bad, right?"

Not bad at all. And then a wave of realization washes over me, and I smile. If I had come here when the water first fell, I would have seen only salt and sand. But this swallow found a home here in the dried-up lake bed. He's made a life in it, and from it. And if the water rises, he'll adapt. And if the water falls again, he'll adapt to that, too. And when his kids are fledged, it'll be their turn to make life from the dried-up lake bed.

And in that fleeting moment, I saw *us*.

And I saw that we were good.

Acknowledgements

I have many, many people to thank for making this book possible. First, a huge thank you to my editors, Jason Kirk and Kai Daniel Lewis, for all of their help in getting my first-ever manuscript into shape. Plenty of people made critical comments and suggestions that shaped the course of this book, including Steve DeDoes, Amara Cohen, Michael Allara, and especially Sapphire Toth, whose "Nanny McPhee" suggestion made all the chapter puzzle pieces finally fit together. I'd like to thank Julie McDowell for her fantastic artwork, and Rachel Hudson, whose incredible cover art gave this book the look it desperately needed. Huge thanks as well to Jonathan Greenberg, for taking on this book and giving me the opportunity of a lifetime.

Special thanks to my mother, Gilda, and my father, Louis, for their endless support on this project. After untold hours spent talking over the phone, suggesting, editing, revising, and re-writing, this book is finally done – and I couldn't have done it without them.

A *Really* Brief History of Time

~13.8 Billion BCE: The Observable Universe is born in a burst of heat, light, and matter.

~13.4 Billion BCE: The first stars are born. The Milky Way begins to form.

~12.8 Billion BCE: The first planets form.

~5 Billion BCE: The Sun is born. It is made from the stellar remnants of a dead ancestor.

~4.6 Billion BCE: The planets of the Solar System form in a rotating planetary disc around the infant Sun. The Earth is a ball of molten rock and is in the midst of the Hadean Bombardment, wrought with comet and asteroid impacts.

~4.2 Billion BCE: The oceans begin to form and reform as Earth's surface is vaporized again and again.

~4.1 Billion BCE: Late Heavy Bombardment. Earth is hit with a constant barrage of asteroids for three hundred million years.

~3.6 Billion BCE: First believed emergence of terrestrial life in the form of simple, single-celled organisms deep within the oceans.

~3.4 Billion BCE: Photosynthetic bacteria evolve and begin to oxygenate Earth's atmosphere.

~2.7 Billion BCE: The first Eukaryotic organisms evolve in a freak accident, as a small bacteria cell fuses with a much larger archaea cell.

~2.3 Billion BCE: Great Oxygenation Event. Oxygen in the atmosphere, toxic to most life on Earth at this time, reaches critical levels and kills 90% of all life.

~1.3 Billion BCE: Multicellular life evolves, and sex evolves soon after.

~600 Million BCE: Fungi evolve.

~540 Million BCE: Cambrian Explosion. Following the mass extinction of the Great Oxygenation Event, complex life bursts onto the scene. Arthropods (which today include insects, spiders, crabs, and shrimp) appear cloaked in exoskeleton, along with the mollusks (which today include squid, octopus, clams, and snails).

~535 Million BCE: First chordates form, the ancestors of any animal with a skeleton. Bony fish swim through the primal oceans.

~500 Million BCE: Plants evolve.

~440 Million BCE: Ordovician-silurian extinction. Many lineages of marine life are snuffed out.

~390 Million BCE: Amphibians evolve from bony fish. They still need to reproduce in water.

~380 Million BCE: The first vertebrates make their way onto land, millions of years after the arthropods.

~365 Million BCE: Devonian Extinction. Even more marine species go extinct.

~359 Million BCE: Carboniferous Period. Earth's land masses are covered in massive swamps and trees, with gigantic insect life. Much of our fossil fuels come from this era.

~320 Million BCE: Reptiles as we know them evolve, using a special amniotic egg that can hatch on land.

~310 Million BCE: Synapsids, the group of reptiles that will eventually evolve into mammals, diverge from Sauropsids, the group that contains modern-day reptiles and birds.

~250 Million BCE: Permian Extinction. The worst mass extinction event in Earth's history wipes out 96% of all marine life and 70% of all life on land.

~240 Million BCE: The first dinosaurs evolve as bipedal predators. They quickly begin to diversify.

~200 Million BCE: Triassic-Jurassic Extinction. The dinosaurs survive and fill ecological niches left open by the deaths of their distant relatives.

~168 Million BCE: Ants evolve.

~66 Million BCE: The reign of the dinosaurs comes to an end as the Cretaceous Extinction wipes most of them from existence. One dinosaur lineage, the birds, manages to survive.

~55 Million BCE: Birds and mammals begin to diversify and fill the ecological niches left by the dinosaurs. The first primates evolve.

~50 Million BCE: Mammals grow larger and larger and continue to diversify. The ancestors of modern whales and dolphins evolve.

~17 Million BCE: The Hominidae, the family that includes Great Apes and modern-day humans, branches off from gibbons and other apes.

~6 Million BCE: The ancestors of modern-day humans and chimpanzees diverge.

~4.4 Million BCE: *Ardipithecus ramidus* evolves, an upright-walking hominid.

~3.5 Million BCE: *Australopithecus afarensis* begins to work with stone tools.

~2.58 Million BCE: The Pleistocene begins. A series of ice ages sweep the Earth, causing mammals to grow giant in size, partly to protect themselves from the cold. Woolly mammoths emerge.

~1 Million BCE: *Homo erectus* creates and uses fire for the very first time.

~450,000 BCE: Neanderthals hunt in the cold regions of modern-day Europe.

~200,000 BCE: The first modern-day humans evolve in Africa. As they migrate to Eurasia, they interbreed with Neanderthals and drive them extinct over the next 150,000 years.

~50,000 BCE: Earliest believed use of human language.

~30,000 BCE: Humans settle the Americas, crossing over a land bridge through Asia.

~12,000 BCE: The Agricultural Revolution. Some humans begin planting crops in fertile river valleys.

~8,000 BCE: The first known cities arise in Mesopotamia.

~4,000 BCE: The Sumerians emerge as one of the world's first civilizations. Civilizations in modern-day Egypt, China, and India soon follow.

~3,200 BCE: The Sumerians invent Cuneiform, the earliest known form of writing.

~2,500 BCE: The Egyptians build the pyramids.

~2,000 BCE: Judaism emerges from a collective of desert-dwelling nomadic herders.

~1,754 BCE: Hammurabi's Code, a set of Babylonian laws, is etched onto a stone obelisk.

~753 BCE: The city of Rome is founded, becoming a republic 200 years later.

~600 BCE: The Buddha is born. Buddhism begins to spread across the East not long after.

~470 BCE: Socrates is born.

~300 BCE: Eratosthenes of Cyrene uses shadows, sticks, and mathematics to demonstrate that the Earth's surface is curved.

~0 CE: Jesus of Nazareth is born.

395 CE: The Roman Empire collapses.

~600 CE: Islam is founded.

1353 CE: The bubonic plague, also known as the Black Death, kills 75-200 million people in Africa and Eurasia.

1492 CE: Christopher Columbus, the genocidal, slave-taking crackpot, makes his famous voyage to the Americas.

1633 YEAR CE: Galileo is placed under house arrest for writing that the earth went around the sun.. 1564 CE: Galileo is born.

1570 CE: Tens of millions die in the Americas as European diseases like smallpox destroy entire civilizations in the northern and southern continents.

1687 CE: Isaac Newton publishes the *Philosiphae Naturalis Principia Mathematica*, paving the way for the emergence of modern science.

1809 CE: Charles Darwin is born in Shrewsbury, England.

1832 CE: Darwin sets sail on the H.M.S. *Beagle*. He will not return to England until 1836.

1859 CE: Darwin publishes *The Origin of Species*, a landmark text for all of biology.

1865 CE: Slavery is abolished in the United States.

1871 CE: Darwin publishes *The Descent of Man*, igniting a firestorm of controversy.

1882 CE: Charles Darwin dies.

1915 CE: Albert Einstein publishes his Theory of General Relativity.

1925 CE: The Scopes Monkey Trial takes place in Dayton, Tennessee.

1997 CE: The author of this book is born. Only he and a few other people consider this to be of any significance.

2008 CE: John Marzluff conducts facial-recognition experiments with crows.

2014 CE: CRISPR is popularized as a genome-editing technology.

2020 CE: The Coronavirus Pandemic hits. The author begins writing this book.

2023 CE: The author "finishes" writing this book.

Author's Note

I've done my best to make this book both readably short, and scientifically accurate. Some concepts, like biological evolution, have so much evidence behind them that they're basically established facts. Other subjects are relatively new, like cultural evolution, which crosses between the fields of biology and anthropology. It's important to distinguish between well established and "new" science, between the deeply grounded and the highly experimental. Both are needed to move us forward.

References

Introduction

1. *English Standard Version Bible*. (2001). Crossway.
2. Darwin Correspondence Project, "Letter no. 2814," accessed on 12 September 2023, https://www.darwinproject.ac.uk/letter/?docId=letters/DCP-LETT-2814.xml
3. Mesoudi, Alex. (2011). *Cultural Evolution: How Darwinian theory can explain human culture and synthesize the social sciences.* Chicago, IL: University of Chicago Press. ISBN 9780226520445.

Chapter 1: Building Moral Compasses

1. Arnhart, L. (1998). *Darwinian Natural Right: The Biological Ethics of Human Nature*. SUNY Press.
2. Boehm, C. (2012). *Moral Origins: The Evolution of Virtue, Altruism, and Shame*. Basic Books.
3. Bradshaw, J. W. S. (2016). Sociality in cats: A comparative review. *Journal of Veterinary Behavior, 11*, 113–124. https://doi.org/10.1016/j.jveb.2015.09.004
4. BRAUBACH, W. (1869). *Religion, Moral & Philosophie der Darwin'schen Artlehre nach ihrer Natur ... Als kleine Parallele menschlich geistiger Entwicklung, etc.*
5. *Code of Hammurabi | Summary & History | Britannica*. (2023, October 10). https://www.britannica.com/topic/Code-of-Hammurabi
6. Darwin, C. (1871). *The Descent of Man: And Selection in Relation to Sex*. Murray.
7. Desmond, A. J., & Moore, J. R. (1994). *Darwin*. W. W. Norton & Company.

8. Lidgard, S., & Love, A. C. (2018). Rethinking Living Fossils. *BioScience, 68*(10), 760–770. https://doi.org/10.1093/biosci/biy084
9. O'Connor, T. (2021). Emergent Properties. In E. N. Zalta (Ed.), *The Stanford Encyclopedia of Philosophy* (Winter 2021). Metaphysics Research Lab, Stanford University. https://plato.stanford.edu/archives/win2021/entries/properties-emergent/
10. Paley, W. (1855). *Natural Theology, Or, Evidences of the Existence and Attributes of the Deity*. Gould and Lincoln.
11. Smay, D., & Armelagos, G. (2000). Galileo Wept: A Critical Assessment of the Use of Race in Forensic Anthropology. *Transforming Anthropology*, 9(2), 19–29. https://doi.org/10.1525/tran.2000.9.2.19

Chapter 2: Cannibalism, a Crash First Course

1. Darwin Correspondence Project, "Letter no. 2814," accessed on 12 September 2023, https://www.darwinproject.ac.uk/letter/?docId=letters/DCP-LETT-2814.xml
2. Schutt, Bill, and Patricia Wynne. *Cannibalism: A Perfectly Natural History*. Algonquin Books of Chapel Hill, 2018.
3. Elgar, Mark A., and Bernard J. Crespi. *Cannibalism: Ecology and Evolution among Diverse Taxa*. Oxford University Press, 1992.
4. Parry, G. D. (1981). The Meanings of r- and K-Selection. *Oecologia*, *48*(2), 260–264.
5. Pfennig, D. W., Reeve, H. K., & Sherman, P. W. (1993). Kin recognition and cannibalism in spadefoot toad tadpoles. *Animal Behaviour*, *46*(1), 87–94. https://doi.org/10.1006/anbe.1993.1164

Chapter 3: Femme Fatale

1. Andrade, M. C. B. (2003). Risky mate search and male self-sacrifice in redback spiders. *Behavioral Ecology*, *14*(4), 531–538. https://doi.org/10.1093/beheco/arg015
2. Courage, K. H. (2014). Octopus!: The Most Mysterious Creature in the Sea. Penguin.
3. Goodall, J. (1977). Infant killing and cannibalism in free-living chimpanzees. *Folia Primatologica; International Journal of Primatology*, *28*(4), 259–289. https://doi.org/10.1159/000155817
4. Hausheer, J. E. (2021, May 24). *Meet the Argonaut, The World's Weirdest Octopus*. Cool Green Science. https://blog.nature.org/science/2021/05/24/meet-the-argonaut-the-worlds-weirdest-octopus/
5. Andrade, M. C. B. (1996). Sexual Selection for Male Sacrifice in the Australian Redback Spider. *Science*, *271*(5245), 70–72. https://doi.org/10.1126/science.271.5245.70
6. Bello, S. M., Saladié, P., Cáceres, I., Rodríguez-Hidalgo, A., & Parfitt, S. A. (2015). Upper Palaeolithic ritualistic cannibalism at Gough's Cave (Somerset, UK): The human remains from head to toe. *Journal of Human Evolution*, *82*, 170–189. https://doi.org/10.1016/j.jhevol.2015.02.016

Chapter 4: Bird eaters, bird feeders

1. Sibley, D. A. (2014). The Sibley Guide to Birds, second edition. Alfred A. Knopf, New York, NY, USA.
2. Koenig, W. D., E. L. Walters, P. B. Stacey, M. T. Stanback, and R. L. Mumme (2019). Acorn Woodpecker (*Melanerpes formicivorus*), version 2.0. In The Birds of North America (P. G. Rodewald, Editor). Cornell Lab of Ornithology, Ithaca, NY, USA. https://doi.org/10.2173/bna.acowoo.02
3. Mumme, R. L., Koenig, W. D., & Pitelka, F. A. (1990). Individual Contributions to Cooperative Nest Care in the Acorn Woodpecker. *The Condor*, *92*(2), 360–368. https://doi.org/10.2307/1368233
4. Koenig, W. D., Mumme, R. L., Stanback, M. T., & Pitelka, F. A. (1995). Patterns and consequences of egg destruction among joint-nesting acorn woodpeckers. *Animal Behaviour*, *50*(3), 607–621. https://doi.org/10.1016/0003-3472(95)80123-5

5. Brode, M., Miller, K. D., Atkins Coleman, A. J., O'Neil, K. L., Poole, L. E., & Bowers, E. K. (2021). Parental favoritism in a wild bird population. *Animal Cognition, 24*(4), 677–687. https://doi.org/10.1007/s10071-020-01463-3

Chapter 5: A Universe of Uneasy Alliances

1. Browne, E. J. (1996). *Charles Darwin: Voyaging.* Princeton University Press.
2. Darwin, C. (1892). *Charles Darwin: His Life Told in an Autobiographical Chapter and in a Selected Series of His Published Letters.* D. Appleton.
3. Darwin, C. (1909). *The Voyage of the Beagle.* P.F. Collier.
4. Herron, M. D., Conlin, P. L., & Ratcliff, W. C. (2022). *The Evolution of Multicellularity.* CRC Press.
5. Phipps, W. E. (2002). *Darwin's Religious Odyssey.* A&C Black.
6. Theobald, D. L. (2010). A formal test of the theory of universal common ancestry. *Nature, 465*(7295), Article 7295. https://doi.org/10.1038/nature09014
7. Payne, J. L., Boyer, A. G., Brown, J. H., Finnegan, S., Kowalewski, M., Krause, R. A., Lyons, S. K., McClain, C. R., McShea, D. W., Novack-Gottshall, P. M., Smith, F. A., Stempien, J. A., & Wang, S. C. (2009). Two-phase increase in the maximum size of life over 3.5 billion years reflects biological innovation and environmental opportunity. *Proceedings of the National Academy of Sciences, 106*(1), 24–27. https://doi.org/10.1073/pnas.0806314106
8. Gray, M. W. (2012). Mitochondrial Evolution. *Cold Spring Harbor Perspectives in Biology,* 4(9), a011403. https://doi.org/10.1101/cshperspect.a011403
9. Roger, A. J., Muñoz-Gómez, S. A., & Kamikawa, R. (2017). The Origin and Diversification of Mitochondria. *Current Biology, 27*(21), R1177–R1192. https://doi.org/10.1016/j.cub.2017.09.015
10. Gray, M. W., Burger, G., & Lang, B. F. (2001). The origin and early evolution of mitochondria. *Genome Biology,* 2(6), reviews1018.1. https://doi.org/10.1186/gb-2001-2-6-reviews1018
11. Cock, J. M. (2021). Evolution of Multicellularity. *Genes, 12*(10), 1532. https://doi.org/10.3390/genes12101532

12. Ursell, L. K., Metcalf, J. L., Parfrey, L. W., & Knight, R. (2012). Defining the Human Microbiome. *Nutrition Reviews*, *70*(Suppl 1), S38–S44. https://doi.org/10.1111/j.1753-4887.2012.00493.x
13. Gilbert, J. A., Blaser, M. J., Caporaso, J. G., Jansson, J. K., Lynch, S. V., & Knight, R. (2018). Current understanding of the human microbiome. *Nature Medicine*, *24*(4), 392–400. https://doi.org/10.1038/nm.4517
14. Grutter, A. S., & Bshary, R. (2003). Cleaner wrasse prefer client mucus: Support for partner control mechanisms in cleaning interactions. *Proceedings of the Royal Society B: Biological Sciences*, *270*(Suppl 2), S242–S244. https://doi.org/10.1098/rsbl.2003.0077
15. Flower, T. P., Gribble, M., & Ridley, A. R. (2014). Deception by Flexible Alarm Mimicry in an African Bird. *Science*, *344*(6183), 513–516. https://doi.org/10.1126/science.1249723
16. Flower, T. (2011). Fork-tailed drongos use deceptive mimicked alarm calls to steal food. *Proceedings of the Royal Society B: Biological Sciences*, *278*(1711), 1548–1555. https://doi.org/10.1098/rspb.2010.1932
17. Flower, T. P., Child, M. F., & Ridley, A. R. (2013). The ecological economics of kleptoparasitism: Pay-offs from self-foraging versus kleptoparasitism. *Journal of Animal Ecology*, *82*(1), 245–255. https://doi.org/10.1111/j.1365-2656.2012.02026.x

Chapter 6: The Ethics of Zombie Ants

1. Browne, E. J. (1996). *Charles Darwin: Voyaging*. Princeton University Press.
2. Darwin, C. (1839). *Journal of Researches Into the Geology and Natural History of the Various Countries Visited by H. M. S. Beagle: Under the Command of Captain FitzRoy, R. N., from 1832 to 1836*. H. Colburn.
3. Desmond, A. J., & Moore, J. R. (1994). *Darwin*. W. W. Norton & Company.
4. Andersen, S. B., Ferrari, M., Evans, H. C., Elliot, S. L., Boomsma, J. J., & Hughes, D. P. (2012). Disease Dynamics in a Specialized Parasite of Ant Societies. *PLOS ONE*, *7*(5), e36352. https://doi.org/10.1371/journal.pone.0036352
5. Bar-On, Y. M., Phillips, R., & Milo, R. (2018). The biomass distribution on Earth. *Proceedings of the National Academy of Sciences*, *115*(25), 6506–6511. https://doi.org/10.1073/pnas.1711842115

6. Browne, E. J. (1996). *Charles Darwin: Voyaging*. Princeton University Press.
7. Conway, J. R. (1986). The Biology of Honey Ants. *The American Biology Teacher*, *48*(6), 335–343. https://doi.org/10.2307/4448321
8. Darwin, C. (1839). *Journal of Researches Into the Geology and Natural History of the Various Countries Visited by H. M. S. Beagle: Under the Command of Captain FitzRoy, R. N., from 1832 to 1836*. H. Colburn.
9. Desmond, A. J., & Moore, J. R. (1994). *Darwin*. W. W. Norton & Company.
10. Eggleton, P. (2020). The State of the World's Insects. *Annual Review of Environment and Resources*, *45*(Volume 45, 2020), 61–82. https://doi.org/10.1146/annurev-environ-012420-050035
11. Fredericksen, M. A., Zhang, Y., Hazen, M. L., Loreto, R. G., Mangold, C. A., Chen, D. Z., & Hughes, D. P. (2017). Three-dimensional visualization and a deep-learning model reveal complex fungal parasite networks in behaviorally manipulated ants. *Proceedings of the National Academy of Sciences*, *114*(47), 12590–12595. https://doi.org/10.1073/pnas.1711673114
12. Giraud, T., Pedersen, J. S., & Keller, L. (2002). Evolution of supercolonies: The Argentine ants of southern Europe. *Proceedings of the National Academy of Sciences*, *99*(9), 6075–6079. https://doi.org/10.1073/pnas.092694199
13. Ivens, A. B., Kronauer, D. J., Pen, I., Weissing, F. J., & Boomsma, J. J. (2012). Ants farm subterranean aphids mostly in single clone groups—An example of prudent husbandry for carbohydrates and proteins? *BMC Evolutionary Biology*, *12*(1), 106. https://doi.org/10.1186/1471-2148-12-106
14. Lu, J. (2019). Cordyceps Zombie Fungus Takes over Ants' Bodies. Accessed June 4, 2022. https://www.nationalgeographic.com/animals/article/cordyceps-zombie-fungus-takes-over-ants.
15. Schultz, T. R. (2000). In search of ant ancestors. *Proceedings of the National Academy of Sciences*, *97*(26), 14028–14029. https://doi.org/10.1073/pnas.011513798
16. Schultz, T. R., & Brady, S. G. (2008). Major evolutionary transitions in ant agriculture. *Proceedings of the National Academy of Sciences*

of the United States of America, *105*(14), 5435–5440. https://doi.org/10.1073/pnas.0711024105

17. Younginger, B. S., Sirová, D., Cruzan, M. B., & Ballhorn, D. J. (2017). Is biomass a reliable estimate of plant fitness? *Applications in Plant Sciences*, *5*(2), 1600094. https://doi.org/10.3732/apps.1600094
18. Zimmer, C. (2019, October 24). After This Fungus Turns Ants Into Zombies, Their Bodies Explode. *The New York Times*. https://www.nytimes.com/2019/10/24/science/ant-zombies-fungus.html

Chapter 7: The Order of Chaos

1. Battisti, C., Poeta, G., & Fanelli, G. (2016). *An Introduction to Disturbance Ecology: A Road Map for Wildlife Management and Conservation*. Springer International Publishing. https://doi.org/10.1007/978-3-319-32476-0
2. Dornelas, M. (2010). Disturbance and change in biodiversity. *Philosophical Transactions of the Royal Society B: Biological Sciences*, *365*(1558), 3719–3727. https://doi.org/10.1098/rstb.2010.0295
3. Dial, R., and Roughgarden, J. (1998). Theory of Marine Communities: The Intermediate Disturbance Hypothesis. *Ecology* 79(4): 1412–24. https://doi.org/10.1890/0012-9658(1998)079[1412:TOMCTI]2.0.CO;2.
4. Climax | Description & Ecology | Britannica. Accessed June 4, 2022. https://www.britannica.com/science/climax-ecology.
5. Weidlich, E. W. A., Nelson, C. R., Maron, J. L., Callaway, R. M., Delory, B. M., & Temperton, V. M. (2021). Priority effects and ecological restoration. *Restoration Ecology*, *29*(1), e13317. https://doi.org/10.1111/rec.13317
6. Connell, J. H. (1961). The Influence of Interspecific Competition and Other Factors on the Distribution of the Barnacle Chthamalus Stellatus. *Ecology*, *42*(4), 710–723. https://doi.org/10.2307/1933500
7. Otto, S. (2008) Sexual Reproduction and the Evolution of Sex. *Nature Education* 1(1):182.

Chapter 8: The Trees That See the Forest

1. Browne, E. J. (1996). *Charles Darwin: Voyaging*. Princeton University Press.
2. Darwin, C. (1839). *Journal of Researches Into the Geology and Natural History of the Various Countries Visited by H. M. S. Beagle: Under the Command of Captain FitzRoy, R. N., from 1832 to 1836*. H. Colburn.
3. Desmond, A. J., & Moore, J. R. (1994). *Darwin*. W. W. Norton & Company.
4. Phipps, W. E. (2002). *Darwin's Religious Odyssey*. A&C Black.
5. Popkin, G. (2019). 'Wood wide web'—The underground network of microbes that connects trees—Mapped for first time. *Science*. https://doi.org/10.1126/science.aay0516
6. Gehring, C., & Bennett, A. (2009). Mycorrhizal Fungal–Plant–Insect Interactions: The Importance of a Community Approach. *Environmental Entomology*, *38*(1), 93–102. https://doi.org/10.1603/022.038.0111
7. Beiler, K. J., Durall, D. M., Simard, S. W., Maxwell, S. A., & Kretzer, A. M. (2010). Architecture of the wood-wide web: Rhizopogon spp. genets link multiple Douglas-fir cohorts. *New Phytologist*, *185*(2), 543–553. https://doi.org/10.1111/j.1469-8137.2009.03069.x
8. Figueiredo, A. F., Boy, J., & Guggenberger, G. (2021). Common Mycorrhizae Network: A Review of the Theories and Mechanisms Behind Underground Interactions. *Frontiers in Fungal Biology*, 2. https://www.frontiersin.org/articles/10.3389/ffunb.2021.735299
9. Simard, S. W., Perry, D. A., Jones, M. D., Myrold, D. D., Durall, D. M., & Molina, R. (1997). Net transfer of carbon between ectomycorrhizal tree species in the field. *Nature*, *388*(6642), Article 6642. https://doi.org/10.1038/41557
10. Teste, F.P., Simard, S.W., Durall, D.M., Guy. R.D., Jones. M.D., and A.L. Schoonmaker. (2009). Access to mycorrhizal networks and tree roots: importance for seedling survival & resource transfer. *Ecology*, 90: 2808-2822.
11. Beiler, K.J., Durall, D.M., Simard, S.W., Maxwell, S.A. and A.M. Kretzer. (2010). Mapping the wood-wide web: mycorrhizal networks link multiple Douglas-fir cohorts. *New Phytologist*, 185: 543-553.

12. Deslippe, J.R., and Simard, S.W. (2011). Below-ground carbon transfer among Betula nana may increase with warming in Arctic tundra. *New Phytologist,* 192(3): 689-698.
13. Simard, S.W., Beiler, K.J., Bingham, M.A., Deslippe. J.R., Philip, L.J., and F.P. Teste. (2012). Mycorrhizal networks: mechanisms, ecology and modelling. Invited Review. *Fungal Biology Reviews,* 26:39-60.
14. Bingham, M.A., and S.W. Simard. (2013). Seedling genetics and life history outweigh mycorrhizal network potential to improve conifer regeneration under drought. *Forest Ecology & Management,* 287:132-139.
15. Simard, S.W., Martin, K., Vyse, A., and Larson, B. (2013). Meta-networks of fungi, fauna and flora as agents of complex adaptive systems. Chapter 7, pages 133-164. *In: Managing World Forests as Complex Adaptive Systems: Building Resilience to the Challenge of Global Change.* Edited by Puettmann, K, Messier, C, and Coates, KD. Routledge, NY. ISBN 978-0-415-51977. 369pp.
16. Beiler, K.J., Simard, S.W. and Durall, D.M. (2015). Topology of *Rhizopogon spp.* mycorrhizal meta-networks in xeric and mesic old-growth interior Douglas-fir forests. *Journal of Ecology, 103(3): 616-628.*
17. Simard, S.W. (2015). Conversations in the forest: The roots of nature's equanimity. *SGI Quarterly,* 79: 8-9.
18. Song, Y.Y. Simard, S.W., Carroll, A., Mohn, W.W. and Zheng, R.S. (2015). Defoliation of interior Douglas-fir elicits carbon transfer and defense signalling to ponderosa pine neighbors through ectomycorrhizal networks. *Nature / Sci. Rep. 5, 8495;* DOI:10.1038/srep08495 (2015).

Chapter 9: The Caws of Culture

1. Browne, E. J. (1996). *Charles Darwin: Voyaging.* Princeton University Press.
2. Browne, E. J. (2011). *Charles Darwin: The Power of Place.* Knopf Doubleday Publishing Group. Chapter 9: The Caws of Culture
3. Darwin, C. (1839). *Journal of Researches Into the Geology and Natural History of the Various Countries Visited by H. M. S. Beagle: Under the Command of Captain FitzRoy, R. N., from 1832 to 1836.* H. Colburn.

4. Darwin, C. (1892). *Charles Darwin: His Life Told in an Autobiographical Chapter and in a Selected Series of His Published Letters*. D. Appleton.
5. Darwin, C. (1909). *The Voyage of the Beagle*. P.F. Collier.
6. Phipps, W. E. (2002). *Darwin's Religious Odyssey*. A&C Black.
7. Brocklehurst, N. (2021). The First Age of Reptiles? Comparing Reptile and Synapsid Diversity, and the Influence of Lagerstätten, During the Carboniferous and Early Permian. *Frontiers in Ecology and Evolution, 9*. https://www.frontiersin.org/articles/10.3389/fevo.2021.669765
8. Seed, A., Emery, N., & Clayton, N. (2009). Intelligence in Corvids and Apes: A Case of Convergent Evolution? *Ethology, 115*(5), 401–420. https://doi.org/10.1111/j.1439-0310.2009.01644.x
9. Pepperberg, I.M. (2019b). Grey parrots: Studies in avian cognition. *Encyclopedia of Animal Behavior, 2nd ed.* J. Choe, ed., Oxford, pp. 83-89.
10. Clements, K. A., Gray, S. L., Gross, B., & Pepperberg, I. M. (2018). Initial evidence for probabilistic reasoning in a grey parrot (Psittacus erithacus). *Journal of Comparative Psychology (Washington, D.C.: 1983), 132*(2), 166–177. https://doi.org/10.1037/com0000106
11. Pepperberg, I.M. (2018a). Grey parrots: Cognitive and communicative abilities. A Practical Guide to Animal Cognition, N. Bueno & F. Amici, eds, Cambridge, UK: Cambridge University Press
12. Waller, B. M., Whitehouse, J., & Micheletta, J. (2017). Rethinking primate facial expression: A predictive framework. *Neuroscience & Biobehavioral Reviews, 82*, 13–21. https://doi.org/10.1016/j.neubiorev.2016.09.005
13. Brusatte et al. (2015). The Origin and Diversification of Birds. Current Biology.
14. Ackerman, Jennifer. (2016). *The Genius of Birds*. Penguin Press.
15. Bock, J., Stein, J., Harnick, S., & Aleichem, S. (1990). *Fiddler on the Roof*. Hal Leonard Corporation.

16. Cornell et al. (2012). Social learning spreads knowledge about dangerous humans among American crows. *Proceedings of the Royal Society B.*
17. Leeuwen, Edwin J. C. van, Katherine A. Cronin, and Daniel B. M. Haun. A Group-Specific Arbitrary Tradition in Chimpanzees (Pan Troglodytes). *Animal Cognition* 17, no. 6 (November 1, 2014): 1421–25. https://doi.org/10.1007/s10071-014-0766-8.
18. Jelbert et al. (2014.) Using the Aesop's Fable Paradigm to Investigate Causal Understanding of Water Displacement by New Caledonian Crows. *PLOS One.*
19. Swift and Marzluff. (2015). Wild American crows gather around their dead to learn about danger. *Journal of Animal Behaviour.*
20. Boeckle, M., & Bugnyar, T. (2012). Long-Term Memory for Affiliates in Ravens. *Current Biology*, 22(9), 801–806. https://doi.org/10.1016/j.cub.2012.03.023
21. Meredith, M. (2009). *Elephant Destiny: Biography Of An Endangered Species In Africa*. PublicAffairs.
22. Moss, Cynthia (1988). *Elephant Memories: Thirteen Years in the Life of an Elephant Family.* University of Chicago Press. ISBN 9780226148533.
23. Boesch, C., Kalan, A. K., Mundry, R., Arandjelovic, M., Pika, S., Dieguez, P., Ayimisin, E. A., Barciela, A., Coupland, C., Egbe, V. E., Eno-Nku, M., Michael Fay, J., Fine, D., Adriana Hernandez-Aguilar, R., Hermans, V., Kadam, P., Kambi, M., Llana, M., Maretti, G., … Kühl, H. S. (2020). Chimpanzee ethnography reveals unexpected cultural diversity. *Nature Human Behaviour, 4*(9), Article 9. https://doi.org/10.1038/s41562-020-0890-1

Chapter 10: Climbing The Cognitive Ladder

1. Sommer, V., Bauer, J., Fowler, A., & Ortmann, S. (2011). Patriarchal Chimpanzees, Matriarchal Bonobos: Potential Ecological Causes of a Pan Dichotomy. In V. Sommer & C. Ross (Eds.), *Primates of Gashaka: Socioecology and Conservation in Nigeria's Biodiversity Hotspot* (pp. 469–501). Springer. https://doi.org/10.1007/978-1-4419-7403-7_12
2. Sapolsky, Robert. (2017). *Behave: The Biology of Humans at Our Best and Worst.* Penguin Publishing.

3. Rapoport, A., Seale, D. A., & Colman, A. M. (2015). Is Tit-for-Tat the Answer? On the Conclusions Drawn from Axelrod's Tournaments. *PLoS ONE, 10*(7), e0134128. https://doi.org/10.1371/journal.pone.0134128
4. Blais, M. J. (1987). Epistemic Tit for Tat. *The Journal of Philosophy, 84*(7), 363–375. https://doi.org/10.2307/2026823
5. Barton, R. A., & Dunbar, R. I. M. (1997). Evolution of the social brain. In A. Whiten & R. W. Byrne (Eds.), *Machiavellian intelligence II: Extensions and evaluations* (pp. 240–263). Cambridge University Press.
6. *English Standard Version Bible.* (2001). Crossway.
7. North, James J. Jr., ""Thou Shalt Not Kill": What Kind of Killing Is Forbidden? Is All Killing Forbidden? (Exodus 20/13; Deuteronomy 5/17)" (2011). *Memory, Meaning & Life.* 98. https://digitalcommons.andrews.edu/mml/98
8. Dunbar, R. I. M. (2003). The Social Brain: Mind, Language, and Society in Evolutionary Perspective. *Annual Review of Anthropology,* 32, 163–181.
9. Boehm, C. (2012). *Moral origins: The evolution of virtue, altruism, and shame.* Basic Books.
10. Carter, G. G., & Wilkinson, G. S. (2013). Food sharing in vampire bats: Reciprocal help predicts donations more than relatedness or harassment. *Proceedings of the Royal Society B: Biological Sciences,* 280(1753), 20122573. https://doi.org/10.1098/rspb.2012.2573
11. Lile, C. (2018, July 11). Science Update: Chimps Start Wars for Power and Benefit From Play. *Jane Goodall's Good for All News.* https://news.janegoodall.org/2018/07/11/chimps-start-wars-power-show-disgust-feces-benefit-play/
12. Richards, R. J. (1987). *Darwin and the Emergence of Evolutionary Theories of Mind and Behavior.* University of Chicago Press.
13. Wade, M. J., & Shuster, S. M. (2004). Sexual Selection: Harem Size and the Variance in Male Reproductive Success. *The American Naturalist, 164*(4), E83–E89. https://doi.org/10.1086/424531
14. Waal, F. de. (2019). *Mama's Last Hug: Animal Emotions and What They Teach Us about Ourselves.* Granta Publications.
15. Martin, D. L. (2021). Violence and Masculinity in Small-Scale Societies. *Current Anthropology, 62*(S23), S169–S181. https://doi.org/10.1086/711689

Chapter 11: A Mind for Others

1. Bowlby, J. (1992). *Charles Darwin: A New Life*. W. W. Norton & Company.
2. Browne, J. (2011). *Charles Darwin: The Power of Place*. Knopf Doubleday Publishing Group.
3. Darwin, C. (2016). *On the Origin of Species*. Xist Publishing.
4. Desmond, A. J., & Moore, J. R. (1994). *Darwin*. W. W. Norton & Company.
5. Phipps, W. E. (2002). *Darwin's Religious Odyssey*. A&C Black.
6. Roberts, J. H. (2010). Religious reactions to Darwin. In P. Harrison (Ed.), *The Cambridge Companion to Science and Religion* (pp. 80–102). Cambridge University Press. https://doi.org/10.1017/CCOL9780521885386.005
7. Vivian Hubert Howard Green. (1996). *A new history of Christianity*. Continuum. http://archive.org/details/newhistoryofchri00gree
8. Sapolsky, R. M. (2017). *Behave: The Biology of Humans at Our Best and Worst*. Penguin.
9. Gallup, G. G., Jr., Anderson, J. R., & Shillito, D. J. (2002). The mirror test. In M. Bekoff, C. Allen, & G. M. Burghardt (Eds.), *The cognitive animal: Empirical and theoretical perspectives on animal cognition* (pp. 325–333). MIT Press
10. Korkiakangas, T., Dindar, K., Laitila, A., & Kärnä, E. (2016). The Sally–Anne test: An interactional analysis of a dyadic assessment. *International Journal of Language & Communication Disorders*, 51(6), 685–702. https://doi.org/10.1111/1460-6984.12240
11. Science Magazine. (2016, October 6). *Humans aren't the only great apes that can 'read minds.'* https://www.youtube.com/watch?v=1sodO_h7q7Q
12. Krupenye, C., Kano, F., Hirata, S., Call, J., & Tomasello, M. (2016). Great apes anticipate that other individuals will act according to false beliefs. *Science*, 354(6308), 110–114. https://doi.org/10.1126/science.aaf8110
13. Langford, D. J., Crager, S. E., Shehzad, Z., Smith, S. B., Sotocinal, S. G., Levenstadt, J. S., Chanda, M. L., Levitin, D. J., & Mogil, J. S. (2006). Social modulation of pain as evidence for empathy in mice. *Science (New York, N.Y.)*, 312(5782), 1967–1970. https://doi.org/10.1126/science.1128322

14. Bartal, I. B.-A., Decety, J., & Mason, P. (2011). Helping a cagemate in need: Empathy and pro-social behavior in rats. *Science (New York, N.Y.)*, *334*(6061), 1427–1430. https://doi.org/10.1126/science.1210789
15. London, Karen. (2021). *Limping Dog Is Empathetic and Intelligent, Not a Con Artist*. The Bark. Retrieved June 6, 2022, from https://thebark.com/content/limping-dog-empathetic-and-intelligent-not-con-artist
16. Hooper, C., Craig, J., Janvrin, D. R., Wetsel, M. A., & Reimels, E. (2010). Compassion Satisfaction, Burnout, and Compassion Fatigue Among Emergency Nurses Compared With Nurses in Other Selected Inpatient Specialties. *Journal of Emergency Nursing*, *36*(5), 420–427. https://doi.org/10.1016/j.jen.2009.11.027
17. Lopez, J., Bindler, R. J., & Lee, J. (2022). Cross-sectional Analysis of Burnout, Secondary Traumatic Stress, and Compassion Satisfaction among Emergency Department Nurses in Southern California Working through the COVID-19 Pandemic. *Journal of Emergency Nursing*, *0*(0). https://doi.org/10.1016/j.jen.2022.03.008
18. Acharya, S., & Shukla, S. (2012). Mirror neurons: Enigma of the metaphysical modular brain. *Journal of Natural Science, Biology, and Medicine*, *3*(2), 118–124. https://doi.org/10.4103/0976-9668.101878
19. Ladygina-Kohts, N. N. (2002). *Infant chimpanzee and human child: A classic 1935 comparative study of ape emotions and intelligence* (pp. xv, 452). Oxford University Press.
20. E.J. Masicampo. (2016, August 31). *A two-year-old's solution to the trolley problem*. https://www.youtube.com/watch?v=-N_RZJUAQY4
21. Sanders, Cheryl. *Lawrence Kohlberg's stages of moral development | Definition & Framework | Britannica*. Retrieved June 6, 2022, from https://www.britannica.com/science/Lawrence-Kohlbergs-stages-of-moral-development

Chapter 12: Empathy Overflowing.

1. Bowlby, J. (1992). *Charles Darwin: A New Life*. W. W. Norton & Company.
2. Browne, J. (2011). *Charles Darwin: The Power of Place*. Knopf Doubleday Publishing Group.
3. Darwin, C., & Burkhardt, F. (1985). *The Correspondence of Charles Darwin: Volume 5, 1851-1855*. Cambridge University Press.

4. Desmond, A. J., & Moore, J. R. (1994). *Darwin*. W. W. Norton & Company.
5. Phipps, W. E. (2002). *Darwin's Religious Odyssey*. A&C Black.
6. Grimm D. (2010). Meeting briefs. Is a dolphin a person?. *Science (New York, N.Y.)*, 327(5969), 1070–1071. https://doi.org/10.1126/science.327.5969.1070-c
7. Bearzi, M. (2012). *Dolphin Confidential: Confessions of a Field Biologist*. University of Chicago Press.
8. BBC Earth. (2021, September 7). *A whale saved my life | Close Encounters | BBC Earth*. https://www.youtube.com/watch?v=OXNCCdcBhcY
9. Pitman, R. L., Deecke, V. B., Gabriele, C. M., Srinivasan, M., Black, N., Denkinger, J., Durban, J. W., Mathews, E. A., Matkin, D. R., Neilson, J. L., Schulman-Janiger, A., Shearwater, D., Stap, P., & Ternullo, R. (2017). Humpback whales interfering when mammal-eating killer whales attack other species: Mobbing behavior and interspecific altruism? *Marine Mammal Science*, *33*(1), 7–58. https://doi.org/10.1111/mms.12343
10. Boehm, C. (2012). *Moral Origins: The Evolution of Virtue, Altruism, and Shame*. Basic Books.
11. Ross, H. M., & Wilson, B. (1996). Violent interactions between bottlenose dolphins and harbour porpoises. *Proceedings of the Royal Society of London. Series B: Biological Sciences*, *263*(1368), 283–286. https://doi.org/10.1098/rspb.1996.0043
12. Patterson, I. A., Reid, R. J., Wilson, B., Grellier, K., Ross, H. M., & Thompson, P. M. (1998). Evidence for infanticide in bottlenose dolphins: An explanation for violent interactions with harbour porpoises? *Proceedings of the Royal Society B: Biological Sciences*, 265(1402), 1167–1170.
13. Harari, Y. N. (2017). *Homo Deus: A Brief History of Tomorrow*. HarperCollins.
14. Rusu, Alina & Krackow, Sven. (2004). Kin-preferential cooperation, dominance-dependent reproductive skew, and competition for mates in communally nesting female house mice. Behavioral Ecology and Sociobiology. 56. 298-305. 10.1007/s00265-004-0787-4.
15. Kaufman, L. (2010, June 16). Ad for a Dish Detergent Becomes Part of a Story. *The New York Times*. https://www.nytimes.com/2010/06/16/science/earth/16dawn.html

16. Entwistle, A., & Dunstone, N. (2000). *Priorities for the Conservation of Mammalian Diversity: Has the Panda Had Its Day?* Cambridge University Press.

Chapter 13: What Darwin Got Wrong.

1. Darwin, C. & Kebler, L. (1859) On the origin of species by means of natural selection, or, The preservation of favoured races in the struggle for life. London: J. Murray. Retrieved from the Library of Congress, https://www.loc.gov/item/06017473/.
2. Browne, J. (2011). *Charles Darwin: The Power of Place*. Knopf Doubleday Publishing Group.
3. Darwin, C. (1871). *The Descent of Man: And Selection in Relation to Sex*. Murray.
4. Desmond, A. J., & Moore, J. R. (1994). *Darwin*. W. W. Norton & Company.
5. Sagan, C. (2011). *Cosmos*. Random House Publishing Group.
6. Schoener, T. W. (1968). The Anolis Lizards of Bimini: Resource Partitioning in a Complex Fauna. *Ecology*, *49*(4), 704–726. https://doi.org/10.2307/1935534
7. "Eratosthenes | Biography, Discoveries, Sieve, & Facts | Britannica." Accessed June 19, 2022. https://www.britannica.com/biography/Eratosthenes.
8. Donihue, C. M., Herrel, A., Fabre, A.-C., Kamath, A., Geneva, A. J., Schoener, T. W., Kolbe, J. J., & Losos, J. B. (2018). Hurricane-induced selection on the morphology of an island lizard. *Nature*, *560*(7716), Article 7716. https://doi.org/10.1038/s41586-018-0352-3
9. Grant, P. R. (1999). *Ecology and Evolution of Darwin's Finches*. Princeton University Press.
10. Grant, P. R., & Grant, B. R. (1995). Predicting Microevolutionary Responses to Directional Selection on Heritable Variation. *Evolution*, *49*(2), 241–251. https://doi.org/10.2307/2410334
11. Grant, P. R., & Grant, B. R. (2002). Unpredictable Evolution in a 30-Year Study of Darwin's Finches. *Science*, *296*(5568), 707–711. https://doi.org/10.1126/science.1070315
12. Hawkins, N. J., Bass, C., Dixon, A., & Neve, P. (2019). The evolutionary origins of pesticide resistance. *Biological Reviews*, *94*(1), 135–155. https://doi.org/10.1111/brv.12440

13. Kleindorfer, S., & Dudaniec, R. Y. (2016). Host-parasite ecology, behavior and genetics: A review of the introduced fly parasite Philornis downsi and its Darwin's finch hosts. *BMC Zoology*, *1*(1), 1. https://doi.org/10.1186/s40850-016-0003-9
14. Losos, J. B. (2018). *Improbable Destinies: Fate, Chance, and the Future of Evolution*. Penguin.
15. Losos, J. B., Warheitt, K. I., & Schoener, T. W. (1997). Adaptive differentiation following experimental island colonization in Anolis lizards. *Nature*, *387*(6628), Article 6628. https://doi.org/10.1038/387070a0
16. Peters, K. J., Evans, C., Aguirre, J. D., & Kleindorfer, S. (2019). Genetic admixture predicts parasite intensity: Evidence for increased hybrid performance in Darwin's tree finches. *Royal Society Open Science*, *6*(4), 181616. https://doi.org/10.1098/rsos.181616
17. Schoener, T. W. (1968). The Anolis Lizards of Bimini: Resource Partitioning in a Complex Fauna. *Ecology*, *49*(4), 704–726. https://doi.org/10.2307/1935534
18. Tompkins, D. M., Mitchell, R. A., & Bryant, D. M. (2006). Hybridization increases measures of innate and cell-mediated immunity in an endangered bird species. *Journal of Animal Ecology*, *75*(2), 559–564. https://doi.org/10.1111/j.1365-2656.2006.01076.x
19. Weiner, J. (1995). *The Beak of the Finch: A Story of Evolution in Our Time (Pulitzer Prize Winner)*. Knopf Doubleday Publishing Group.
20. Centers for Disease Control and Prevention (U.S.) (2019). Antibiotic resistance threats in the United States, 2019.

Chapter 14: Trials, Tribulations, and Unintentional Invasions.

1. Brun, A., Magallanes, M. E., Barrett-Wilt, G. A., Karasov, W. H., & Caviedes-Vidal, E. (2021). Dietary adaptation to high starch involves increased relative abundance of sucrase-isomaltase and its mRNA in nestling house sparrows. *American Journal of Physiology-Regulatory, Integrative and Comparative Physiology*, *320*(2), R195–R202. https://doi.org/10.1152/ajpregu.00181.2020
2. Jukar, A. M., Lyons, S. K., Wagner, P. J., & Uhen, M. D. (2021). Late Quaternary extinctions in the Indian Subcontinent. *Palaeogeography, Palaeoclimatology, Palaeoecology*, *562*, 110137. https://doi.org/10.1016/j.palaeo.2020.110137

3. Ravinet, M., Elgvin, T. O., Trier, C., Aliabadian, M., Gavrilov, A., & Sætre, G.-P. (2018). Signatures of human-commensalism in the house sparrow genome. *Proceedings of the Royal Society B: Biological Sciences, 285*(1884), 20181246. https://doi.org/10.1098/rspb.2018.1246
4. Remmert, H. (1982). The evolution of man and the extinction of animals. *Naturwissenschaften, 69*(11), 524–527. https://doi.org/10.1007/BF00463496
5. Sætre, G.-P., Riyahi, S., Aliabadian, M., Hermansen, J. S., Hogner, S., Olsson, U., Gonzalez Rojas, M. F., Sæther, S. A., Trier, C. N., & Elgvin, T. O. (2012). Single origin of human commensalism in the house sparrow. *Journal of Evolutionary Biology, 25*(4), 788–796. https://doi.org/10.1111/j.1420-9101.2012.02470.x
6. Sheldon, E. L., & Griffith, S. C. (2017). A high incidence of non-cavity nesting in an introduced population of House Sparrows suggests that the species should not be constrained by cavity-nest site availability. *Avian Research, 8*(1), 29. https://doi.org/10.1186/s40657-017-0087-0

Chapter 15: The Men Who Shoot at Goats

1. Abrantes, J., van der Loo, W., Le Pendu, J., & Esteves, P. J. (2012). Rabbit haemorrhagic disease (RHD) and rabbit haemorrhagic disease virus (RHDV): A review. *Veterinary Research, 43*(1), 12. https://doi.org/10.1186/1297-9716-43-12
2. Carrion, V., Donlan, C. J., Campbell, K. J., Lavoie, C., & Cruz, F. (2011). Archipelago Wide Island Restoration in the Galápagos Islands: Reducing Costs of Invasive Mammal Eradication Programs and Reinvasion Risk. *PLOS ONE, 6*(5), e18835. https://doi.org/10.1371/journal.pone.0018835
3. Jolly, C. J., Shine, R., & Greenlees, M. J. (2015). The impact of invasive cane toads on native wildlife in southern Australia. *Ecology and Evolution, 5*(18), 3879–3894. https://doi.org/10.1002/ece3.1657
4. Palmer, J. (2021, March 29). *Fire as Medicine: Learning from Native American Fire Stewardship*. Eos. http://eos.org/features/fire-as-medicine-learning-from-native-american-fire-stewardship

5. Project Isabela. (2023). *Galápagos Conservancy*. Retrieved November 18, 2023, from https://www.galapagos.org/conservation/project-isabela/
6. Ripple, W. J., & Beschta, R. L. (2012). Trophic cascades in Yellowstone: The first 15 years after wolf reintroduction. *Biological Conservation, 145*(1), 205–213. https://doi.org/10.1016/j.biocon.2011.11.005
7. Schellhase, J. (2015, February 28). *Project Isabela: When Slaughtering 250,000 Goats Meant Saving A Species*. All That's Interesting. https://allthatsinteresting.com/project-isabela
8. Shine, R., Ward-Fear, G., & Brown, G. P. (2020). A famous failure: Why were cane toads an ineffective biocontrol in Australia? *Conservation Science and Practice*, 2(12), e296. https://doi.org/10.1111/csp2.296

Chapter 16: The Making of Homo Sapiens

1. Browne, J. (2011). *Charles Darwin: The Power of Place*. Knopf Doubleday Publishing Group.
2. Darwin, C. (1871). *The Descent of Man: And Selection in Relation to Sex*. Murray.
3. Darwin, C. (1872). *The expression of the emotions in man and animals*. John Murray. https://doi.org/10.1037/10001-000
4. Desmond, A. J., & Moore, J. R. (1994). *Darwin*. W. W. Norton & Company.
5. Ekman, P. (2009). Darwin's contributions to our understanding of emotional expressions. *Philosophical Transactions of the Royal Society B: Biological Sciences, 364*(1535), 3449–3451. https://doi.org/10.1098/rstb.2009.0189
6. Austad, S. N. (1994). Menopause: An evolutionary perspective. *Experimental Gerontology*, 29(3), 255–263. https://doi.org/10.1016/0531-5565(94)90005-1
7. Boehm, C. (2012). *Moral Origins: The Evolution of Virtue, Altruism, and Shame*. Basic Books.
8. Chagnon, N. A. (2012). *The Yanomamo*. Cengage Learning.
9. De Stefani, E., & De Marco, D. (2019). Language, Gesture, and Emotional Communication: An Embodied View of Social Interaction. *Frontiers in Psychology, 10*. https://www.frontiersin.org/articles/10.3389/fpsyg.2019.02063

10. Dunbar, R. I. M. (1992). Neocortex size as a constraint on group size in primates. *Journal of Human Evolution, 22*(6), 469–493. https://doi.org/10.1016/0047-2484(92)90081-J
11. Garcia, R. (2017). Neurobiology of fear and specific phobias. *Learning & Memory*, 24(9), 462–471. https://doi.org/10.1101/lm.044115.116
12. *Homo heidelbergensis | The Smithsonian Institution's Human Origins Program*. Retrieved November 18, 2023, from http://humanorigins.si.edu/evidence/human-fossils/species/homo-heidelbergensis
13. Huxley, A. (2010). *Brave New World* (11th ed.). Vintage.
14. Ko, K. H. (2016). Hominin interbreeding and the evolution of human variation. *Journal of Biological Research-Thessaloniki, 23*(1), 17. https://doi.org/10.1186/s40709-016-0054-7
15. Nattrass, S., Croft, D. P., Ellis, S., Cant, M. A., Weiss, M. N., Wright, B. M., Stredulinsky, E., Doniol-Valcroze, T., Ford, J. K. B., Balcomb, K. C., & Franks, D. W. (2019). Postreproductive killer whale grandmothers improve the survival of their grandoffspring. *Proceedings of the National Academy of Sciences, 116*(52), 26669–26673. https://doi.org/10.1073/pnas.1903844116
16. Noonan, J. P. (2010). Neanderthal genomics and the evolution of modern humans. *Genome Research*, 20(5), 547–553. https://doi.org/10.1101/gr.076000.108
17. O'Connor, S., Ono, R., & Clarkson, C. (2011). Pelagic Fishing at 42,000 Years Before the Present and the Maritime Skills of Modern Humans. *Science*, 334(6059), 1117–1121. https://doi.org/10.1126/science.1207703
18. Pollycove, R., Naftolin, F., & Simon, J. A. (2011). The evolutionary origin and significance of menopause. *Menopause, 18*(3), 336. https://doi.org/10.1097/gme.0b013e3181ed957a
19. Reader, S. M., Hager, Y., & Laland, K. N. (2011). The evolution of primate general and cultural intelligence. *Philosophical Transactions of the Royal Society B: Biological Sciences, 366*(1567), 1017–1027. https://doi.org/10.1098/rstb.2010.0342
20. Sandler, W. (2018). The Body as Evidence for the Nature of Language. *Frontiers in Psychology, 9*. https://www.frontiersin.org/articles/10.3389/fpsyg.2018.01782

21. Schulkin, J. (2016). Evolutionary Basis of Human Running and Its Impact on Neural Function. *Frontiers in Systems Neuroscience, 10*. https://www.frontiersin.org/articles/10.3389/fnsys.2016.00059
22. Shepherd, S. (2010). Following Gaze: Gaze-Following Behavior as a Window into Social Cognition. *Frontiers in Integrative Neuroscience, 4*. https://www.frontiersin.org/articles/10.3389/fnint.2010.00005
23. Shermer, M. (2005). *The Science of Good and Evil: Why People Cheat, Gossip, Care, Share, and Follow the Golden Rule*. Macmillan.
24. Takahata, N., & Satta, Y. (1997). Evolution of the primate lineage leading to modern humans: Phylogenetic and demographic inferences from DNA sequences. *Proceedings of the National Academy of Sciences, 94*(9), 4811–4815. https://doi.org/10.1073/pnas.94.9.4811
25. Tomasello, M., Hare, B., Lehmann, H., & Call, J. (2007). Reliance on head versus eyes in the gaze following of great apes and human infants: The cooperative eye hypothesis. *Journal of Human Evolution*, 52(3), 314–320. https://doi.org/10.1016/j.jhevol.2006.10.001
26. Waal, F. de. (2019). *Mama's Last Hug: Animal Emotions and What They Tell Us about Ourselves*. W. W. Norton & Company.
27. Wang, Z., Quinn, P. C., Tanaka, J. W., Yu, X., Sun, Y.-H. P., Liu, J., Pascalis, O., Ge, L., & Lee, K. (2015). An other-race effect for configural and featural processing of faces: Upper and lower face regions play different roles. *Frontiers in Psychology, 6*. https://www.frontiersin.org/articles/10.3389/fpsyg.2015.00559

Chapter 17: Monkey See, Monkey Considers Moral Implications of Situation

1. Bartholomew, M. J., Jones, R. V., & Paton, W. D. M. (1997). The Award of the Copley Medal to Charles Darwin. *Notes and Records of the Royal Society of London, 30*(2), 209–218. https://doi.org/10.1098/rsnr.1976.0014
2. Browne, J. (2011). *Charles Darwin: The Power of Place*. Knopf Doubleday Publishing Group.
3. Burkhardt, F. H. (2001). Darwin and the Copley Medal. *Proceedings of the American Philosophical Society, 145*(4), 510–518.
4. Darwin, C. (2001). *The Correspondence of Charles Darwin*. 12. 1864. United Kingdom: Cambridge University Press.

5. TED Blog Video (Director). (2013, April 4). *Two Monkeys Were Paid Unequally: Excerpt from Frans de Waal's TED Talk.* https://www.youtube.com/watch?v=meiU6TxysCg
6. van Wolkenten, M., Brosnan, S. F., & de Waal, F. B. M. (2007). Inequity responses of monkeys modified by effort. *Proceedings of the National Academy of Sciences, 104*(47), 18854–18859. https://doi.org/10.1073/pnas.0707182104
7. Sapolsky, Robert. (2017). Behave: The Biology of Humans at Our Best and Worst. Penguin Publishing.
8. Brosnan, S. F., & de Waal, F. B. M. (2014). Evolution of responses to (un)fairness. *Science (New York, N.Y.), 346*(6207), 1251776. https://doi.org/10.1126/science.1251776

Chapter 18: A Moral Evolution

1. Bregman, R. (2020). *Humankind: A Hopeful History*. Little, Brown.
2. Bruner, J. (1991). The Narrative Construction of Reality. *Critical Inquiry, 18*(1), 1–21.
3. Creanza, N., Kolodny, O., & Feldman, M. W. (2017). Cultural evolutionary theory: How culture evolves and why it matters. *Proceedings of the National Academy of Sciences, 114*(30), 7782–7789. https://doi.org/10.1073/pnas.1620732114
4. Darwin, C. (1871). *The Descent of Man. And Selection in Relation to Sex*. Murray.
5. Grandin, G. (2010). *Fordlandia: The Rise and Fall of Henry Ford's Forgotten Jungle City*. Henry Holt and Company.
6. Harari, Y. N. (2017). *Homo Deus: A Brief History of Tomorrow*. HarperCollins.
7. Lewens, T. (2020). Cultural Evolution. In E. N. Zalta (Ed.), *The Stanford Encyclopedia of Philosophy* (Summer 2020). Metaphysics Research Lab, Stanford University. https://plato.stanford.edu/archives/sum2020/entries/evolution-cultural/
8. Lopez-Minguez, J., Morosoli, J. J., Madrid, J. A., Garaulet, M., & Ordoñana, J. R. (2017). Heritability of siesta and night-time sleep as continuously assessed by a circadian-related integrated measure. *Scientific Reports*, 7, 12340. https://doi.org/10.1038/s41598-017-12460-x

9. Magazine, S., & Mott, J. M., Justin. (2016). *A Journey to the Oldest Cave Paintings in the World*. Smithsonian Magazine. Retrieved July 3, 2022, from https://www.smithsonianmag.com/history/journey-oldest-cave-paintings-world-180957685/
10. Mesoudi, A. (2011). *Cultural Evolution: How Darwinian Theory Can Explain Human Culture and Synthesize the Social Sciences*. University of Chicago Press.
11. Richerson, P. J., & Boyd, R. (2008). *Not By Genes Alone: How Culture Transformed Human Evolution*. University of Chicago Press.
12. Singelis, T. M., Triandis, H. C., Bhawuk, D. P. S., & Gelfand, M. J. (1995). Horizontal and Vertical Dimensions of Individualism and Collectivism: A Theoretical and Measurement Refinement. *Cross-Cultural Research*, *29*(3), 240–275. https://doi.org/10.1177/106939719502900302
13. Talhelm, T., Zhang, X., Oishi, S., Shimin, C., Duan, D., Lan, X., & Kitayama, S. (2014). Large-Scale Psychological Differences Within China Explained by Rice Versus Wheat Agriculture. *Science*, *344*(6184), 603–608. https://doi.org/10.1126/science.1246850
14. Tubaro, P., Ryan, L., & D'angelo, A. (2016). The Visual Sociogram in Qualitative and Mixed-Methods Research. *Sociological Research Online*, *21*(2), 180–197. https://doi.org/10.5153/sro.3864
15. Van de Vliert, E. (2007). Climates Create Cultures. *Social and Personality Psychology Compass*, *1*(1), 53–67. https://doi.org/10.1111/j.1751-9004.2007.00003.x

Chapter 19: Are You There, Darwin? It's Me, God.

1. Brown, F. B. (1986). The Evolution of Darwin's Theism. *Journal of the History of Biology*, *19*(1), 1–45.
2. Browne, J. (2011). *Charles Darwin: The Power of Place*. Knopf Doubleday Publishing Group.
3. Darwin, C. (1892). *Charles Darwin: His Life Told in an Autobiographical Chapter and in a Selected Series of His Published Letters*. D. Appleton.
4. Darwin, C. (1993). *The correspondence of Charles Darwin. 8. 1860*. Cambridge University Press.
5. Darwin, C. (1871). *The Descent of Man: And Selection in Relation to Sex*. Murray.

6. Desmond, A. J., & Moore, J. R. (1994). *Darwin*. W. W. Norton & Company.
7. New International Version. (1984). International Bible Society. (Original work published 1973).
8. Arnold, M. (1883). *Literature & dogma*. Macmillan.
9. Banerji, U. (500, 00:00). *Lemurs Get High on Their Millipede Supply*. Atlas Obscura. http://www.atlasobscura.com/articles/lemurs-get-high-on-their-millipede-supply
10. Barker, G. (2006). *The Agricultural Revolution in Prehistory: Why did Foragers become Farmers?* OUP Oxford.
11. Broom, D. M. (2003). *The Evolution of Morality and Religion*. Cambridge University Press.
12. Harris, R. (2010, December 23). Did 'Shrooms Send Santa And His Reindeer Flying? *NPR*. https://www.npr.org/2010/12/24/132260025/did-shrooms-send-santa-and-his-reindeer-flying
13. Held, A. (2018, October 4). Minnesota Residents Call Police On Rowdy Drunk Birds. *NPR*. https://www.npr.org/2018/10/04/654489250/minnesota-residents-call-police-on-rowdy-drunk-birds
14. Irwin, C., Leveritt, M., Shum, D., & Desbrow, B. (2013). The effects of dehydration, moderate alcohol consumption, and rehydration on cognitive functions. *Alcohol (Fayetteville, N.Y.)*, *47*(3), 203–213. https://doi.org/10.1016/j.alcohol.2012.12.016
15. Jamison, S., & Brereton, J. (2020). *The Rigveda: A Guide*. Oxford University Press.
16. Koenig, H. G. (2012). Religion, Spirituality, and Health: The Research and Clinical Implications. *ISRN Psychiatry*, *2012*, 278730. https://doi.org/10.5402/2012/278730
17. Koole, S. L., Meijer, M., & Remmers, C. (2017). Religious rituals as tools for adaptive self-regulation. *Religion, Brain & Behavior*, *7*(3), 250–253. https://doi.org/10.1080/2153599X.2016.1156562
18. Lewis-Williams, D., & Pearce, D. (2005). *Inside the Neolithic Mind: Consciousness, Cosmos, and the Realm of the Gods*. Thames & Hudson.
19. Lindenfors, P., & Svensson, J. (2021). Evolutionary explanations for religion: An interdisciplinary critical review. *Research Ideas and Outcomes*, *7*, e66132. https://doi.org/10.3897/rio.7.e66132

20. McKay, R., & Whitehouse, H. (2015). Religion and Morality. *Psychological Bulletin, 141*(2), 447–473. https://doi.org/10.1037/a0038455
21. Morris, B. J., Krieger, J. N., & Klausner, J. D. (2017). CDC's Male Circumcision Recommendations Represent a Key Public Health Measure. *Global Health: Science and Practice*, 5(1), 15–27. https://doi.org/10.9745/GHSP-D-16-00390
22. Nielsen, M., Langley, M. C., Shipton, C., & Kapitány, R. (2020). Homo neanderthalensis and the evolutionary origins of ritual in Homo sapiens. *Philosophical Transactions of the Royal Society B: Biological Sciences*, 375(1805), 20190424. https://doi.org/10.1098/rstb.2019.0424
23. Purzycki, B. G., & Sosis, R. (2009). The Religious System as Adaptive: Cognitive Flexibility, Public Displays, and Acceptance. In E. Voland & W. Schiefenhövel (Eds.), *The Biological Evolution of Religious Mind and Behavior* (pp. 243–256). Springer. https://doi.org/10.1007/978-3-642-00128-4_17
24. Religion. (2022). In *Wikipedia*. https://en.wikipedia.org/w/index.php?title=Religion&oldid=1095665820
25. Roberts, K. E. (1963). Mechanism of Dehydration Following Alcohol Ingestion. *Archives of Internal Medicine, 112*(2), 154–157. https://doi.org/10.1001/archinte.1963.03860020052002
26. Schoenherr, R. A. (1987). Power and Authority in Organized Religion: Disaggregating the Phenomenological Core. *Sociological Analysis*, 47, 52–71. https://doi.org/10.2307/3711652
27. Staff, H. H. P. (2018, June 29). *Intermittent fasting: The positive news continues*. Harvard Health. https://www.health.harvard.edu/blog/intermittent-fasting-surprising-update-2018062914156
28. Trepanowski, J. F., & Bloomer, R. J. (2010). The impact of religious fasting on human health. *Nutrition Journal*, 9(1), 57. https://doi.org/10.1186/1475-2891-9-57
29. Willard, A. K., Baimel, A., Turpin, H., Jong, J., & Whitehouse, H. (2020). Rewarding the good and punishing the bad: The role of karma and afterlife beliefs in shaping moral norms. *Evolution and Human Behavior*, 41(5), 385–396. https://doi.org/10.1016/j.evolhumbehav.2020.07.001

Chapter 20: A Will To Change

1. Browne, J. (2011). *Charles Darwin: The Power of Place*. Knopf Doubleday Publishing Group.
2. BBC Earth. (2009, November 6). *The Intense 8 Hour Hunt | Attenborough Life of Mammals | BBC Earth*. https://www.youtube.com/watch?v=826HMLoiE_0
3. Boehm, C. (2012). *Moral Origins: The Evolution of Virtue, Altruism, and Shame*. Basic Books.
4. Cooper, G. H. (1987.). *COYOTE IN NAVAJO RELIGION AND COSMOLOGY*. 13.
5. Darwin, C. (1892). *Charles Darwin: His Life Told in an Autobiographical Chapter and in a Selected Series of His Published Letters*. D. Appleton.
6. Darwin, C. (1911). *The Life and Letters of Charles Darwin: Including an Autobiographical Chapter*. D. Appleton.
7. Davis, E. F. (2008). *Scripture, Culture, and Agriculture: An Agrarian Reading of the Bible*. Cambridge University Press.
8. Desmond, A. J., & Moore, J. R. (1994). *Darwin*. W. W. Norton & Company.
9. Erdoes, R., & Ortiz, A. (1985). *American Indian Myths and Legends*. Knopf Doubleday Publishing Group.
10. Reimers, Mark. (2014). *Ethics Series*. Retrieved July 3, 2022, from http://www.people.vcu.edu/~mreimers/Personal/early.human.ethics.html
11. Flannery, K. V. (2019). *The Early Mesoamerican Village: Updated Edition*. Routledge.
12. Harari, Y. N. (2017). *Homo Deus: A Brief History of Tomorrow*. HarperCollins.
13. Hawkes, K., O'Connell, J., & Blurton Jones, N. (2018). Hunter-gatherer studies and human evolution: A very selective review. *American Journal of Physical Anthropology*, *165*(4), 777–800. https://doi.org/10.1002/ajpa.23403
14. Judson, K. B. (1913). *Myths and Legends of the Great Plains*. A.C. McClurg & Company.
15. Marean, C. W. (2016). The transition to foraging for dense and predictable resources and its impact on the evolution of modern

humans. *Philosophical Transactions of the Royal Society B: Biological Sciences, 371*(1698), 20150239. https://doi.org/10.1098/rstb.2015.0239

16. Matthews, W. (1886). Navajo Names for Plants. *The American Naturalist, 20*(9), 767–777.
17. Migliano, A. B., & Vinicius, L. (2022). The origins of human cumulative culture: From the foraging niche to collective intelligence. *Philosophical Transactions of the Royal Society B: Biological Sciences, 377*(1843), 20200317. https://doi.org/10.1098/rstb.2020.0317
18. Mosse, D. (2018). Caste and development: Contemporary perspectives on a structure of discrimination and advantage. *World Development, 110*, 422–436. https://doi.org/10.1016/j.worlddev.2018.06.003
19. Lewis, Orrin. *Native American Indian Buffalo Legends, Meaning and Symbolism from the Myths of Many Tribes.* (2020). Retrieved July 3, 2022, from http://www.native-languages.org/legends-buffalo.htm
20. Johns, Elizabeth Louise. (2020). Buffalo Renaissance: The Northern Plains Tribes' Path to Self-Determination. *Graduate Student Theses, Dissertations, & Professional Papers. 11639.* https://scholarworks.umt.edu/etd/11639
21. Peoples, H. C., Duda, P., & Marlowe, F. W. (2016). Hunter-Gatherers and the Origins of Religion. *Human Nature (Hawthorne, N.y.), 27*, 261–282. https://doi.org/10.1007/s12110-016-9260-0
22. Sankaran, S., Sekerdej, M., & von Hecker, U. (2017). The Role of Indian Caste Identity and Caste Inconsistent Norms on Status Representation. *Frontiers in Psychology, 8*, 487. https://doi.org/10.3389/fpsyg.2017.00487
23. Smith, A. (1776). *An Inquiry Into the Nature and Causes of the Wealth of Nations*. Whitestone.

Chapter 21: Driving, Genes, and Gene Drives

1. Browne, J. (2011). *Charles Darwin: The Power of Place*. Knopf Doubleday Publishing Group.
2. Darwin, C. (1911). *The Life and Letters of Charles Darwin: Including an Autobiographical Chapter*. D. Appleton.
3. CDC. (2022, June 7). *What is Sickle Cell Disease? | CDC*. Centers for Disease Control and Prevention. https://www.cdc.gov/ncbddd/sicklecell/facts.html

4. *China scientist condemned for gene-editing babies, loses job*. (2019). Al Jazeera. Retrieved November 19, 2023, from https://www.aljazeera.com/news/2019/1/22/china-scientist-condemned-for-gene-editing-babies-loses-job
5. Gewin, V. (2003). Genetically Modified Corn—Environmental Benefits and Risks. *PLoS Biology*, *1*(1), e8. https://doi.org/10.1371/journal.pbio.0000008
6. Hammond, A., Pollegioni, P., Persampieri, T., North, A., Minuz, R., Trusso, A., Bucci, A., Kyrou, K., Morianou, I., Simoni, A., Nolan, T., Müller, R., & Crisanti, A. (2021). Gene-drive suppression of mosquito populations in large cages as a bridge between lab and field. *Nature Communications*, *12*(1), Article 1. https://doi.org/10.1038/s41467-021-24790-6
7. Hansson, S. O., Belin, M.-Å., & Lundgren, B. (2021). Self-Driving Vehicles—An Ethical Overview. *Philosophy & Technology*, *34*(4), 1383–1408. https://doi.org/10.1007/s13347-021-00464-5
8. Janssens, A. C. J. W. (2016). Designing babies through gene editing: Science or science fiction? *Genetics in Medicine*, *18*(12), Article 12. https://doi.org/10.1038/gim.2016.28
9. Kallioinen, N., Pershina, M., Zeiser, J., Nosrat Nezami, F., Pipa, G., Stephan, A., & König, P. (2019). Moral Judgements on the Actions of Self-Driving Cars and Human Drivers in Dilemma Situations From Different Perspectives. *Frontiers in Psychology*, *10*. https://www.frontiersin.org/article/10.3389/fpsyg.2019.02415
10. Khamsi, R. (2020). Gene therapy could offer an inclusive cure for cystic fibrosis. *Nature*, *583*(7818), S12–S14. https://doi.org/10.1038/d41586-020-02111-z
11. Ledford, H. (2020). Quest to use CRISPR against disease gains ground. *Nature*, *577*(7789), 156. https://doi.org/10.1038/d41586-019-03919-0
12. Levy, N. (2004). *What Makes Us Moral?* Oneworld Publications.
13. Li, J., Walker, S., Nie, J., & Zhang, X. (2019). Experiments that led to the first gene-edited babies: The ethical failings and the urgent need for better governance. *Journal of Zhejiang University. Science. B*, *20*(1), 32–38. https://doi.org/10.1631/jzus.B1800624
14. Lino, C. A., Harper, J. C., Carney, J. P., & Timlin, J. A. (2018). Delivering CRISPR: A review of the challenges and approaches.

Drug Delivery, 25(1), 1234–1257.
https://doi.org/10.1080/10717544.2018.1474964
15. Min, J., Smidler, A. L., Najjar, D., & Esvelt, K. M. (2018). Harnessing gene drive. *Journal of Responsible Innovation*, 5(sup1), S40–S65. https://doi.org/10.1080/23299460.2017.1415586
16. Mitra, S., Behbahani, H., & Eriksdotter, M. (2019). Innovative Therapy for Alzheimer's Disease-With Focus on Biodelivery of NGF. *Frontiers in Neuroscience, 13*, 38. https://doi.org/10.3389/fnins.2019.00038
17. Pellegrino, E., Bedini, S., Nuti, M., & Ercoli, L. (2018). Impact of genetically engineered maize on agronomic, environmental and toxicological traits: A meta-analysis of 21 years of field data. *Scientific Reports*, *8*(1), Article 1. https://doi.org/10.1038/s41598-018-21284-2
18. Regalado, A. (2019). *China's CRISPR babies: Read exclusive excerpts from the unseen original research.* MIT Technology Review. Retrieved July 3, 2022, from https://www.technologyreview.com/2019/12/03/131752/chinas-crispr-babies-read-exclusive-excerpts-he-jiankui-paper/
19. Rodríguez-Rodríguez, D. R., Ramírez-Solís, R., Garza-Elizondo, M. A., De Lourdes Garza-Rodríguez, M., & Barrera-Saldaña, H. A. (2019). Genome editing: A perspective on the application of CRISPR/Cas9 to study human diseases (Review). *International Journal of Molecular Medicine*, *43*(4), 1559–1574. https://doi.org/10.3892/ijmm.2019.4112
20. Schmerge, Alexandra. (2018). Designer Babies: A Paired Analysis of the Technological Advances and Ethical Implications of Genetic Selection. *Honor Scholar Theses, 92*. https://scholarship.depauw.edu/studentresearch/92
21. Stepanichev, M. (2020). Gene Editing and Alzheimer's Disease: Is There Light at the End of the Tunnel? *Frontiers in Genome Editing*, 2. https://www.frontiersin.org/articles/10.3389/fgeed.2020.00004
22. Treffert, D. A. (2009). The savant syndrome: An extraordinary condition. A synopsis: past, present, future. *Philosophical Transactions of the Royal Society B: Biological Sciences*, *364*(1522), 1351–1357. https://doi.org/10.1098/rstb.2008.0326

23. Wang, H., & Yang, H. (2019). Gene-edited babies: What went wrong and what could go wrong. *PLOS Biology, 17*(4), e3000224. https://doi.org/10.1371/journal.pbio.3000224
24. Yan, S. (2018, November 28). China's "Dr Frankenstein" says second woman in early pregnancy with gene-edited babies. *The Telegraph.* https://www.telegraph.co.uk/news/2018/11/28/chinas-dr-frankenstein-says-second-woman-early-pregnancy-gene/

Chapter 22: Adaptation, Across the Nation

1. Abrevaya, X. C., Leitzinger, M., Oppezzo, O. J., Odert, P., Patel, M. R., Luna, G. J. M., Forte Giacobone, A. F., & Hanslmeier, A. (2020). The UV surface habitability of Proxima b: First experiments revealing probable life survival to stellar flares. *Monthly Notices of the Royal Astronomical Society: Letters, 494*(1), L69–L74. https://doi.org/10.1093/mnrasl/slaa037
2. Anderson, C. (2021). *The Second: Race and Guns in a Fatally Unequal America.* Bloomsbury Publishing USA.
3. Ayala, F. J. (2010). The difference of being human: Morality. *Proceedings of the National Academy of Sciences, 107*(supplement_2), 9015–9022. https://doi.org/10.1073/pnas.0914616107
4. Beauvoir, S. D. (2012). *The Second Sex.* Knopf Doubleday Publishing Group.
5. Bible proves that life begins at conception. (2019). *Lewistownsentinel.com.* Retrieved July 3, 2022, from https://www.lewistownsentinel.com/opinion/letters-to-the-editor/2019/11/bible-proves-that-life-begins-at-conception/
6. Buchanan, M. (2017). The risks of accelerated change. *Nature Physics, 13*(9), Article 9. https://doi.org/10.1038/nphys4249
7. Darwin, C. (1871). *The Descent of Man, and Selection in Relation to Sex: In Two Volumes.* Murray.
8. Gandhi, M., Gandhi, M. K., & Attenborough, R. (1982). *The Words of Gandhi.* Newmarket Press.
9. Howard, M. A. (2022, July 20). *What the Bible actually says about abortion may surprise you.* The Conversation. http://theconversation.com/what-the-bible-actually-says-about-abortion-may-surprise-you-186983

10. Kamimura, Y., Yoshizawa, K., Lienhard, C., Ferreira, R. L., & Abe, J. (2021). Evolution of nuptial gifts and its coevolutionary dynamics with male-like persistence traits of females for multiple mating. *BMC Ecology and Evolution, 21*(1), 164. https://doi.org/10.1186/s12862-021-01901-x
11. McLennan, D. A. (2008). The Concept of Co-option: Why Evolution Often Looks Miraculous. *Evolution: Education and Outreach, 1*(3), Article 3. https://doi.org/10.1007/s12052-008-0053-8
12. New International Version. (1984). International Bible Society. (Original work published 1973).
13. Schenker, J. G. (2008). The beginning of human life. *Journal of Assisted Reproduction and Genetics, 25*(6), 271–276. https://doi.org/10.1007/s10815-008-9221-6
14. Senter, P., & Moch, J. G. (2015). A critical survey of vestigial structures in the postcranial skeletons of extant mammals. *PeerJ, 3*, e1439. https://doi.org/10.7717/peerj.1439
15. Shermer, M. (2005). *The Science of Good and Evil: Why People Cheat, Gossip, Care, Share, and Follow the Golden Rule*. Macmillan.

Chapter 23: Reconciling Morality With Evolutionary Biology

1. Browne, J. (2011). *Charles Darwin: The Power of Place*. Knopf Doubleday Publishing Group.
2. Crooks, K. R., Gonzalez, M. N., & Ghasemi, B. (2022). Scientist perspectives toward the status and management of gray wolves in the western United States. *Conservation Science and Practice*, 4(7), e12728. https://doi.org/10.1111/csp2.12728
3. Darwin, C. (1911). *The Life and Letters of Charles Darwin: Including an Autobiographical Chapter*. D. Appleton.
4. Underwood, B. F. (1882). *The Index: A Weekly Paper*. Index association.
5. Anslow, L. (2023). In 1903, New York Times predicted that airplanes would take 10 million years to develop. *Big Think*. Retrieved July 3, 2022, from https://bigthink.com/pessimists-archive/air-space-flight-impossible/
6. Appell, D. (2008). *The Sun Will Eventually Engulf Earth—Maybe*. Scientific American. Retrieved July 3, 2022, from https://www.scientificamerican.com/article/the-sun-will-eventually-engulf-earth-maybe/

7. Bonanno, G. A. (2004). Loss, Trauma, and Human Resilience: Have We Underestimated the Human Capacity to Thrive After Extremely Aversive Events? *American Psychologist*, 59(1), 20–28. https://doi.org/10.1037/0003-066X.59.1.20
8. Bregman, R. (2020a). *Humankind: A Hopeful History*. Little, Brown.
9. Bregman, R. (2020b, May 9). The real Lord of the Flies: What happened when six boys were shipwrecked for 15 months. *The Guardian*. https://www.theguardian.com/books/2020/may/09/the-real-lord-of-the-flies-what-happened-when-six-boys-were-shipwrecked-for-15-months
10. Britt, C. L. (1997). Reconsidering the unemployment and crime relationship: Variation by age group and historical period. *Journal of Quantitative Criminology*, *13*(4), 405–428. https://doi.org/10.1007/BF02221048
11. Cajori, F. (1908). *The Age of the Sun and the Earth*. Scientific American. https://doi.org/10.1038/scientificamerican09121908-174supp
12. Darwin, C. (1871). *The Descent of Man, and Selection in Relation to Sex: In Two Volumes*. Murray.
13. Frost, Robert (1923). *Fire and Ice*. New Hampshire.
14. Graham, C. (2021, February 10). America's crisis of despair: A federal task force for economic recovery and societal well-being. *Brookings*. https://www.brookings.edu/research/americas-crisis-of-despair-a-federal-task-force-for-economic-recovery-and-societal-well-being/
15. *Great Green Wall Initiative*. (2022). UNCCD. Retrieved July 3, 2022, from https://www.unccd.int/our-work/ggwi
16. Hawks, J., Hunley, K., Lee, S.-H., & Wolpoff, M. (2000). Population Bottlenecks and Pleistocene Human Evolution. *Molecular Biology and Evolution*, *17*(1), 2–22. https://doi.org/10.1093/oxfordjournals.molbev.a026233
17. King, M. (2018). *The Good Neighbor: The Life and Work of Fred Rogers*. Abrams.
18. Krulwich, R. (2012, October 22). How Human Beings Almost Vanished From Earth In 70,000 B.C. *NPR*. https://www.npr.org/sections/krulwich/2012/10/22/163397584/how-human-beings-almost-vanished-from-earth-in-70-000-b-c
19. Lockwood, G. W., Skiff, B. A., Baliunas, S. L., & Radick, R. R. (1992). Long-term solar brightness changes estimated from a survey of Sun-

like stars. *Nature, 360*(6405), Article 6405. https://doi.org/10.1038/360653a0

20. Meltzer, D. J. (2020). Overkill, glacial history, and the extinction of North America's Ice Age megafauna. *Proceedings of the National Academy of Sciences, 117*(46), 28555–28563. https://doi.org/10.1073/pnas.2015032117
21. News, S. S., E&E. (2021). *Military Operations Will be Strained by Climate Change*. Scientific American. Retrieved July 3, 2022, from https://www.scientificamerican.com/article/military-operations-will-be-strained-by-climate-change/
22. Osborne, M. (2022). *After 35 Years of Recovery Efforts, Bald Eagles Are No Longer Considered Endangered in Vermont | Smart News| Smithsonian Magazine*. Retrieved July 3, 2022, from https://www.smithsonianmag.com/smart-news/bald-eagle-removed-from-vermonts-endangered-species-list-180979582/
23. Prialnik, D. (2010). *An introduction to the theory of stellar structure and evolution* (2nd ed). Cambridge University Press.
24. Raphael, S., & Winter-Ebmer, R. (2001). Identifying the Effect of Unemployment on Crime. *The Journal of Law & Economics, 44*(1), 259–283. https://doi.org/10.1086/320275
25. Rodríguez, H., Trainor, J., & Quarantelli, E. L. (2006). Rising to the Challenges of a Catastrophe: The Emergent and Prosocial Behavior following Hurricane Katrina. *The Annals of the American Academy of Political and Social Science, 604*(1), 82–101. https://doi.org/10.1177/0002716205284677
26. Schwartz, B. (2022, July 12). *Jan. 6 committee notifies DOJ that Trump tried tampering with one of its witnesses, Cheney says*. CNBC. https://www.cnbc.com/2022/07/12/jan-6-committee-notifies-doj-that-trump-tried-tampering-with-one-of-its-witnesses-cheney-says.html
27. Smith, D. W., & Peterson, R. O. (2021). Intended and unintended consequences of wolf restoration to Yellowstone and Isle Royale National Parks. *Conservation Science and Practice, 3*(4), e413. https://doi.org/10.1111/csp2.413
28. Assistant Secretary of the Army (2021). US Army: Addressing Climate Change Threats. Retrieved July 3, 2022, from https://www.army.mil/standto/archive/2021/05/14/

29. Szalavitz, M. (2011, March 22). 'Paradise Built in Hell:' How Disaster Brings Out the Best in People. *Time*. https://healthland.time.com/2011/03/22/a-paradise-built-in-hell-how-disaster-brings-out-the-best-in-people/
30. Szalavitz, M. (2012, October 31). How Disasters Bring Out Our Kindness. *Time*. https://healthland.time.com/2012/10/31/how-disasters-bring-out-our-kindness/
31. Tangermann, V. (2019). *Modelling galactic settlement*. Retrieved July 3, 2022, from https://www.esa.int/ESA_Multimedia/Images/2019/06/Modelling_galactic_settlement
32. Valdivia, A., Wolf, S., & Suckling, K. (2019). Marine mammals and sea turtles listed under the U.S. Endangered Species Act are recovering. *PLOS ONE*, *14*(1), e0210164. https://doi.org/10.1371/journal.pone.0210164
33. Wadsworth, T. P. (2001). *Employment, Crime, and Context: A Multi-Level Analysis of the Relationship between Work and Crime*. 244.
34. Wehrl, A. (1978). General properties of entropy. *Reviews of Modern Physics*, 50(2), 221–260. https://doi.org/10.1103/RevModPhys.50.221
35. *World War II bombing associated with resilience, not 'German Angst.'* (2017, June 23). University of Cambridge. https://www.cam.ac.uk/research/news/world-war-ii-bombing-associated-with-resilience-not-german-angst
36. Zerbini, A. N., Adams, G., Best, J., Clapham, P. J., Jackson, J. A., & Punt, A. E. (2019). Assessing the recovery of an Antarctic predator from historical exploitation. *Royal Society Open Science*, *6*(10), 190368. https://doi.org/10.1098/rsos.190368

Chapter 24: What's the Problem With Evil?

1. Boehm, C. (2012). *Moral Origins: The Evolution of Virtue, Altruism, and Shame*. Basic Books.
2. Bregman, R. (2020). *Humankind: A Hopeful History*. Little, Brown.
3. Burns, J. M., & Swerdlow, R. H. (2003). Right Orbitofrontal Tumor With Pedophilia Symptom and Constructional Apraxia Sign. *Archives of Neurology*, *60*(3), 437–440. https://doi.org/10.1001/archneur.60.3.437

4. Chang, I. (2014). *The Rape Of Nanking: The Forgotten Holocaust Of World War II*. Basic Books.
5. Cikara, M., Botvinick, M. M., & Fiske, S. T. (2011). Us versus Them: Social Identity Shapes Neural Responses to Intergroup Competition and Harm. *Psychological Science, 22*(3), 10.1177/0956797610397667. https://doi.org/10.1177/0956797610397667
6. Darwin, C. (1871). *The Descent of Man, and Selection in Relation to Sex: In Two Volumes*. Murray.
7. Darwin, C. (1908). *Charles Darwin: His Life Told in an Autobiographical Chapter, and in a Selected Series of His Published Letters*. John Murray.
8. Flynn, F. J., & Lake, V. K. B. (2008). If you need help, just ask: Underestimating compliance with direct requests for help. *Journal of Personality and Social Psychology, 95*(1), 128–143. https://doi.org/10.1037/0022-3514.95.1.128
9. Inc, G. (2016, September 20). *Billions Worldwide Help Others in Need*. Gallup.Com. https://news.gallup.com/poll/195659/billions-worldwide-help-others-need.aspx
10. Kozlowska, K., Walker, P., McLean, L., & Carrive, P. (2015). Fear and the Defense Cascade: Clinical Implications and Management. *Harvard Review of Psychiatry, 23*(4), 263–287. https://doi.org/10.1097/HRP.0000000000000065
11. Kunimatsu, M., & Marsee, M. (2012). Examining the Presence of Anxiety in Aggressive Individuals: The Illuminating Role of Fight-or-Flight Mechanisms. *Child & Youth Care Forum, 41*. https://doi.org/10.1007/s10566-012-9178-6
12. Levy, N. (2004). *What Makes Us Moral?* Oneworld Publications.
13. Palix, J., Gillespie, S. M., Abbiati, M., & Abu-Akel, A. (2022). Dynamics of parasympathetic activity in violent incarcerated offenders before, during, and in recovery from an emotional inhibition task. *Scientific Reports, 12*(1), Article 1. https://doi.org/10.1038/s41598-022-10872-y
14. Ponge, J.-F. (2005). Emergent properties from organisms to ecosystems: Towards a realistic approach. *Biological Reviews of the Cambridge Philosophical Society, 80*(3), 403–411. https://doi.org/10.1017/S146479310500672X
15. Sapolsky, R. M. (2017). *Behave: The Biology of Humans at Our Best and Worst*. Penguin.

16. Shermer, M. (2005). *The Science of Good and Evil: Why People Cheat, Gossip, Care, Share, and Follow the Golden Rule*. Macmillan.
17. Toland, J. (2003). *The Rising Sun: The Decline and Fall of the Japanese Empire, 1936-1945*. Random House Publishing Group.

Chapter 25: The Voyage Ahead

1. Wiens, J. A., Patten, D. T., & Botkin, D. B. (1993). Assessing Ecological Impact Assessment: Lessons from Mono Lake, California. *Ecological Applications*, 3(4), 595–609. https://doi.org/10.2307/1942093
2. Sahagun, L., & James, I. (2022, April 15). *As drought hammers Mono Lake, thirsty Los Angeles must look elsewhere for water*. Los Angeles Times. https://www.latimes.com/california/story/2022-04-15/l-a-gets-less-water-from-mono-lake-due-to-declining-levels
3. *Mono Lake Basin | California State Water Resources Control Board*. (2022). Retrieved July 4, 2022, from https://www.waterboards.ca.gov/waterrights/water_issues/programs/mono_lake/

Made in the USA
Monee, IL
13 September 2026

63137521R00187